Asian Cobras

This book comprehensively explores Asian Cobras, covering their evolutionary history, geographic distribution, venom characteristics, pathophysiology of envenomation, and treatment strategies. The introductory chapters focus on venomous snakes, placing particular emphasis on cobras and the substantial impact of snake envenomation in Asia, where it is recognized as a neglected tropical disease. The venom delivery system, including glands, dentition, and fangs, is explored in detail. The book also explores their appearance, behavior, habitat, feeding, and mating habits, along with their geographic distribution across Asia. Further, the chapters examine the venom composition of different cobra species, highlighting the variations within and between species as well as across geographic regions, supported by advanced biochemical and proteomic analyses. The medical consequences of cobra envenomation are examined, focusing on clinical manifestations such as neurotoxicity, cytotoxicity, and other serious complications, while emphasizing the role of venom variability in determining the severity of envenomation.

Asian Cobras

Venom Evolution, Composition, Pharmacology and Treatment

Ashis Kumar Mukherjee
Bhargab Kalita

CRC Press
Taylor & Francis Group
Boca Raton New York London

CRC Press is an imprint of the
Taylor & Francis Group, an **informa** business

Designed cover image: Shutterstock

First edition published 2025
by CRC Press
2385 NW Executive Center Drive, Suite 320, Boca Raton FL 33431

and by CRC Press
4 Park Square, Milton Park, Abingdon, Oxon, OX14 4RN

CRC Press is an imprint of Taylor & Francis Group, LLC

British Library Cataloguing-in-Publication Data
A catalogue record for this book is available from the British Library

ISBN: 978-1-032-98331-8 (hbk)
ISBN: 978-1-032-98332-5 (pbk)
ISBN: 978-1-003-59812-1 (ebk)

DOI: 10.1201/9781003598121

Typeset in Times
by SPi Technologies India Pvt Ltd (Straive)

Contents

About the Authors

Prof. Ashis Kumar Mukherjee is currently serving as the Director of the Institute of Advanced Study in Science and Technology (I.A.S.S.T.), Guwahati. He completed his M.Sc. in Biochemistry from Banaras Hindu University in 1992, his Ph.D. in 1998 on Indian cobra and Russell's viper venoms from the Department of Biochemistry, Burdwan Medical College, Burdwan University, and his D.Sc. in Biotechnology from Calcutta University in 2017, focusing on snake venom phospholipase A_2 and protease enzymes. With more than 30 years of research experience in characterizing snake and scorpion venom toxins, his current research interests include proteomic analysis of snake and scorpion venoms, pharmacological re-assessment of medicinal plants used against snakebite and scorpion stings, quality assessment of commercial antivenoms, and cardiovascular and anticancer drug discovery from natural resources, including snake and scorpion venoms. Prof. Mukherjee has received several awards and medals for his academic and research achievements, including the prestigious Visitor's Award for Research in Basic and Applied Sciences from the Honorable President of India in 2018. He is a fellow of the Royal Society of Biology, U.K., the West Bengal Academy of Science and Technology, Kolkata, India, the Indian Academy of Science, Bangalore, the National Academy of Science, India, Prayagraj, and the National Academy of Medical Sciences, New Delhi. He has published over 175 research articles in peer-reviewed international and national journals, as well as book chapters and review articles. Additionally, he serves as a task force member for the D.B.T., I.C.M.R., and WHO on the prevention and treatment of snakebite envenoming.

Dr. Bhargab Kalita is currently serving as an Assistant Professor at the Amrita Research Center, Faridabad. He completed his M.Sc. in Biotechnology from St. Anthony's College in 2013 and earned his Ph.D. in 2019 from Tezpur University, specializing in the proteomics and antivenomics of Russell's viper venoms. With over five years of experience in venom research, Dr. Kalita has published more than 30 research articles in peer-reviewed international journals.

1 Introduction

1.1 INTRODUCTION

The limbless reptiles known as snakes or serpents are categorized under the suborder Serpentes. They are thought to have evolved from terrestrial lizards (order Squamata) between 70 and 140 million years ago during the Cretaceous period (McDowell, 1972), and they flourished during the Neogene period, which is aptly dubbed the 'Age of Snakes' (Stanley, 1979). Snakes are often classified into two groups: venomous and non-venomous, depending on their capacity to secrete venom, a highly specialized toxic substance. Except for Antarctica and a few islands, such as Greenland, Iceland, and Ireland, there are over 3000 species of snakes worldwide, only about 15% of which are venomous (Fox and Serrano, 2008; Mebs, 2002).

All venomous creatures, including snakes, contain venom glands, which are primarily responsible for synthesizing and preserving a variety of toxins for immediate release and protecting the snake from its venom (Mackessy, 2002; Mackessy and Baxter, 2006; Munekiyo and Mackessy, 1998). Numerous proteins and peptides found in snake venom have evolved rapidly through positive Darwinian evolution to help immobilize, kill, and digest prey (Casewell et al., 2013; Daltry et al., 1996; Fry, 2005; Ohno et al., 1997). These zootoxins are very selective and have a high affinity toward the neuromuscular and cardiovascular systems and the blood coagulation of the snakebite victim or prey (Mukherjee et al., 2000; Mukherjee, 2014; Mukherjee and Mackessy, 2013; Mukherjee and Maity, 2002; Thakur and Mukherjee, 2017; Warrell, 1999; White, 2005). However, with careful characterization, this range of medically significant toxins also shows promise as potential life-saving medications (Kalita et al., 2021; Koh and Kini, 2012; Mukherjee and Mackessy, 2013; Thakur and Mukherjee, 2017).

Approximately 600 species of snakes belonging to the four families Colubridae, Viperidae, Elapidae, and Atractaspididae are known to be venomous out of the 3000 species that are found across the world (Fox and Serrano, 2008; Mackessy, 2002; Mackessy and Saviola, 2016). With about 1,800 species, Colubridae is the most prominent family of snakes, and its members are found all over the world (Hill and Mackessy, 2000; Junqueira-de-Azevedo et al., 2015; Mackessy, 2002; Mackessy and Saviola, 2016). Seven subfamilies of Colubridae comprise many medically significant species, including Boomslang (*Dispholidus typus*), Brown Treesnake (*Boiga irregularis*), and Red-necked Keel-back (*Rhabdophis subminiatus*), among others (Lawson et al., 2005; Smith and Yeo, 1977). Viperinae (true vipers belonging to genus *Bitis*, *Cerastes*, *Echis*, and *Daboia*), Azemiopinae (with only one recognized member, *Azemiops feae*), and Crotalinae (pit vipers belonging to *Crotalus*, *Agkistrodon*, and *Bothrops* genera) comprise the family Viperidae, which includes about 310 species. Primarily terrestrial, these venomous viperid snakes can be found

DOI: 10.1201/9781003598121-1

throughout Asia, Africa, Europe, and America (Fox and Serrano, 2008; Lawson et al., 2005; Smith and Yeo, 1977).

The Elapidae family has about 350 species spread across Africa, America, the Middle East, Asia, and Australasia. The family is further subdivided into two subfamilies: Elapinae (belonging to genus *Naja*, *Bungarus*, *Dendroaspis*, *Oxyuranus*, *Calliophis*, and *Micrurus*) and Hydrophiinae (sea snakes of genus *Laticauda*, *Enhydrina*, *Hydrophis*, and *Microcephalophis*) (Lawson et al., 2005; Smith and Yeo, 1977). Only a few snakes with significant medical value are found in the Atractaspididae family, distributed throughout Africa and the Middle East. These snakes include the southern African stiletto (*Atractaspis bibronii*) and the burrowing asp (*Atractaspis engaddensis*) (Akani et al., 2001; Smith and Yeo, 1977).

It is noteworthy that all four families of snakes have venoms that contain a variety of toxin classes, some of which have significant sequence homology among them. Nonetheless, venoms from each family have unique biological characteristics due to notable variations in amino acid sequence and relative toxin content (Fox and Serrano, 2008; Mackessy, 2010; Mackessy and Saviola, 2016). For instance, the venoms of the Viperidae family are typically abundant in enzymatic proteins that impact the victims' or preys' hemostatic system (Kalita et al., 2018; Kalita et al., 2017; Mukherjee et al., 2000; Mukherjee et al., 2016; Patra et al., 2017; Tan et al., 2015). Conversely, the venom of the Elapidae family of snakes is highly concentrated in neurotoxic proteins (Ali et al., 2013; Chanda et al., 2019; Chanda and Mukherjee, 2020; Chanda et al., 2018; Dutta et al., 2017; Viala et al., 2015; Wong et al., 2018). Additionally, the venoms of Atractaspididae family of snakes contain peptide toxins that have an impact on the blood vascular system (Fox and Serrano, 2008; Shine et al., 2006), and Colubridae venoms exhibit clinical manifestations similar to both viperid and elapid envenomation (Peichoto et al., 2012; Weldon and Mackessy, 2010).

1.2 THE SNAKEBITE PROBLEM – A NEGLECTED TROPICAL DISEASE

An estimated 5 million people encounter snake envenomation annually, and nearly 100,000 people succumb to death (Chippaux, 1998). According to the most recent epidemiological study on snake envenoming, which was conducted in 227 countries, over 421,000 people worldwide are bitten by venomous snakes every year, resulting in approximately 20,000 deaths due to venomous snakebite (Kasturiratne et al., 2008). However, snakebite deaths may reach 94,000 yearly because many countries do not routinely report snakebite cases (Kasturiratne et al., 2008). Moreover, few nations have an accurate epidemiological reporting system for snakebite data (Chippaux, 1998; Gutierrez et al., 2010; Kasturiratne et al., 2008). As a result, it is difficult to determine the actual worldwide incidence of snake envenoming, fatality rate, and related problems, including morbidity, due to a lack of data, which is startling and frightening (Gutierrez et al., 2010; Kasturiratne et al., 2008). An approach for a current assessment of the worldwide incidence of snakebite was developed through a literature review on snake envenoming and modeling based on geographical projections of snakebite and mortality (Gutierrez et al., 2010; Kasturiratne et al., 2008). The authors' conclusion, based on their data, is that snakebite-related morbidity and mortality represent a serious global health risk, with Southeast Asia, South Asia, and sub-Saharan Africa bearing the brunt of the snakebite incidence.

Snake envenoming is a critical occupational health issue that predominantly affects the rural and agricultural workers of developing nations in Oceania, Africa, Asia, and Latin America (Figure 1.1, Table 1.1) (Gutierrez et al., 2017; Kasturiratne et al., 2008; WHO, 2007). According to the conservative estimates, the highest incidence of snakebite occurs in South Asia (121,000 cases), followed by Southeast Asia (111,000 cases) and East Sub-Saharan Africa (43,000 cases) (Alirol et al., 2010; Gutierrez et al., 2006; Kasturiratne et al., 2008). Furthermore, in the world, the highest number of snakebite envenomations occurs in India (81,000 yearly), followed by Sri Lanka (33,000), Vietnam (30,000), Brazil (30,000), Mexico (28,000), Pakistan (20,000), and Nepal (20,000), according to the report presented by Kasturiratne et al. (2008). In addition, according to a recent study, there are over 314,000 snakebite cases in sub-Saharan Africa each year, which lead to 6,000 amputations and 7,300 fatalities (Chippaux et al., 2011). Still, West Africa does not recognize or prioritize the issue of snakebite (Habib et al., 2015). The high prevalence of morbidity and mortality following snake envenomation in these areas can be attributed to two significant factors: a deficient healthcare system and a shortage of antivenom (Mukherjee, 2021; Mukherjee and Mackessy, 2021; Warrell, 2013).

Despite the emotional, psychological, and socioeconomic repercussions, regional and national health authorities worldwide have not prioritized snakebite prevention, treatment, or systematic research on snake venoms. However, in 2011, the World Health Organization (WHO) recognized snakebite as a neglected tropical illness, and since then, global interest and understanding of this severe health concern has expanded. It is important to stress that developing, standardizing, and optimizing protocols for reporting snakebite incidents should be the top priority. Furthermore, to tackle this significant health concern, it is of essential importance to preserve reliable records of the morbidity and death from snake envenomation in medical facilities, particularly in developing countries (Kasturiratne et al., 2008; Mathers et al., 2007; Mukherjee, 2021).

1.3 THE IMPACT OF SNAKEBITE IN DEVELOPING NATIONS

In Europe, snakebite cases are rarely life-threatening, although they can occasionally result in serious complications (Chippaux, 2012). The incidence (per 100,000 population) of snake envenoming in Europe ranges from 0.09 to 2.09, while the mortality rate (per 100,000 population) ranges between 0.01 and 0.03 (Kasturiratne et al., 2008). According to reports in 2012 by Chippaux, there were 7992 incidents of snake envenoming throughout Europe (including European Russia and Turkey); however, only 15% of the bites were considered severe and may lead to death or disability (Chippaux, 2012). Every year in Great Britain, about 200 persons are admitted to hospitals as a result of snakebite; nevertheless, since 1975, there have been no recorded deaths. On the other hand, France has a relatively higher rate of snakebite incidents than the United Kingdom (Warrell, 1996).

According to Chippaux et al. (1995), the Department of Yonne (150 km south of Paris) reported five snakebite cases per 100,000 residents annually, and the rate is

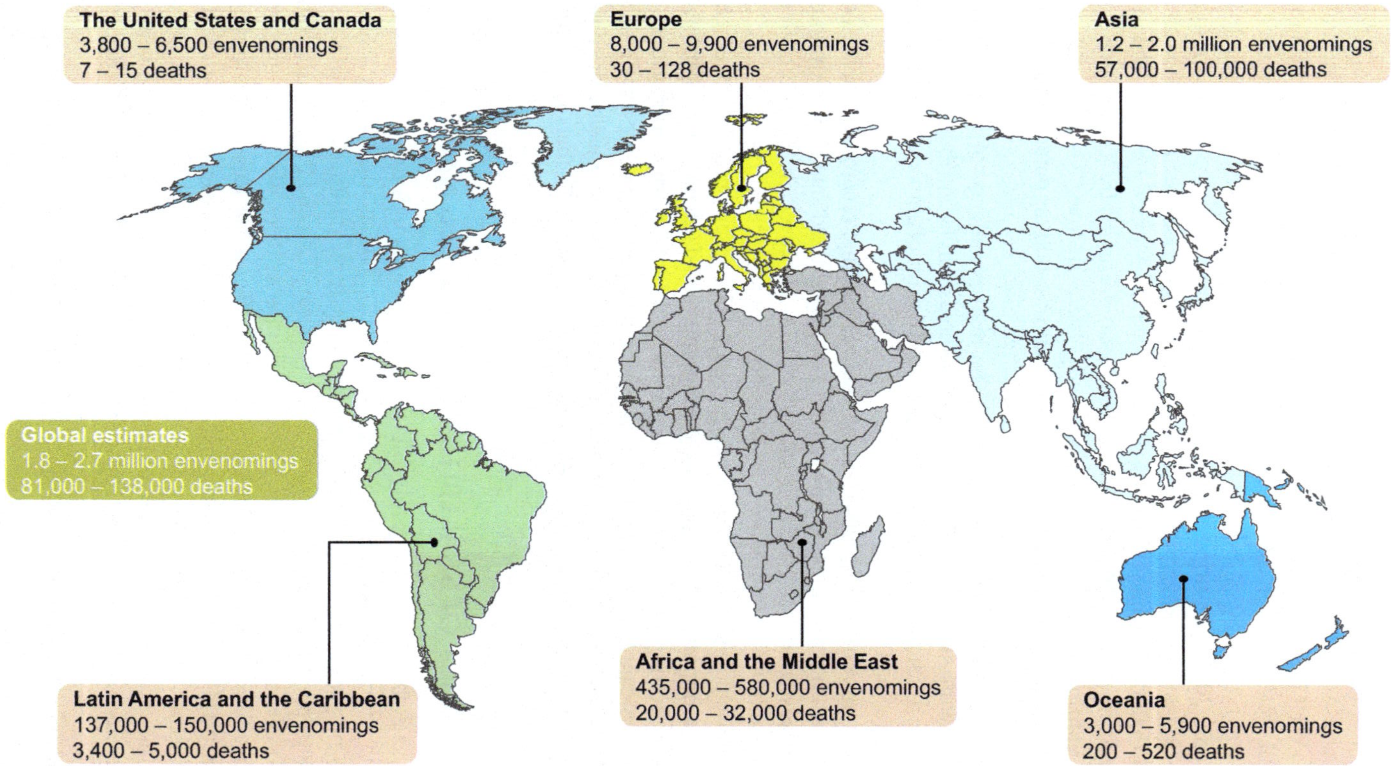

FIGURE 1.1 Geographical distribution of the estimated number of snakebite envenomings and deaths.

(Adapted from the original picture of Gutierrez et al., 2017.)

TABLE 1.1
Estimated Snakebite Cases and Mortality in Different Regions of the World

Region	Snakebite Estimate (High and Low Ranges)	Annual Incidence/100,000	Snakebite Mortality (High and Low Ranges)	Annual Mortality/100,000
East Asia	4,582–218,673	0.33–15.73	462–4,829	0.033–0.347
South Asia	121,333–463,350	7.84–29.94	14,112–33,666	0.912–2.175
Southeast Asia	110,533–498,158	18.82–84.65	790–19,094	0.134–3.245
Australasia	1,099–1260	4.41–5.06	2–4	0.008–0.016
Central Europe	106–2,489	0.09–2.09	6–17	0.005–0.014
Eastern Europe	795–795	0.38–0.38	37–66	0.018–0.032
Western Europe	3,060–6,618	0.74–1.61	5–45	0.001–0.011
Andean Latin America	6,548–27,653	12.90–54.47	243–533	0.479–1.050
Central Latin America	42,087–67,373	19.04–30.47	193–1,461	0.087–0.661
Southern Latin America	2,058–2,163	3.46–3.63	4–5	0.007–0.008
Tropical Latin America	29,636–31,895	14.97–16.12	100–299	0.051–0.151
North Africa/Middle East	3,017–80,191	0.71–18.88	43–78	0.010–0.018
North America	2,683–3,858	0.79–1.14	5–7	0.001–0.002
Oceania	361–4,635	3.87–49.70	227–516	2.434–5.533
Central Sub-Saharan Africa	18,176–47,820	20.28–53.37	256–3,083	0.286–3.441
Eastern Sub-Saharan Africa	42,834–74,823	12.94–22.61	1,397–10,014	0.422–3.026
Southern Sub-Saharan Africa	1,613–2,296	2.34–3.33	286–366	0.414–0.530
Western Sub-Saharan Africa	27,999–294,700	8.87–93.34	1,590–18,654	0.504–5.909

Source: Adapted from Kasturiratne et al. (2008).

comparable in other parts of the country (Chippaux, 1998). Among other venomous species distributed in France, envenomation by *Vipera berus* and *Vipera aspis* can be fatal and necessitates prompt medical attention (de Haro, 2012). Rural areas in Southern Europe, including those in Italy and Spain, have also been reported with higher incidence of morbidity from snakebites (5 per 100,000 persons). However, Switzerland has relatively lower rates of snake envenomation; per 100,000 people, there are only 0.1 cases of envenomation annually (Stahel et al., 1985).

The annual rate of snake envenoming in the United States and Canada is comparable to that of Europe. Approximately 45,000 snakebites are reported annually in North America; however, the mortality rate is relatively low (15 deaths out of 45,000 envenomations), which is relatively low due to better snakebite treatment facilities (Da Silva et al., 2003). However, snakebite incidents are far more common in Central and South America. Between 1990 and 2000, about 5,550 snakebite incidents were reported in Costa Rica (Sasa and Vazquez, 2003). The majority of the victims in these snakebite cases were rural residents, particularly agricultural laborers. Notably, the Central American lancehead (*Bothrops asper*) is the primary culprit responsible for Costa Rica's high rate of snakebite fatalities (Sasa and Vazquez, 2003).

Australia has significantly fewer snakebite occurrences; 300 cases yearly require medical attention, with 1–4 cases resulting in fatalities (Stewart, 2003). However, compared to the decreased frequency of snake envenomation in temperate regions of Australia, snakebite mortality and morbidity throughout tropical Australia seem substantial (Cheng and Currie, 2004).

1.4 THE IMPACT OF SNAKEBITE IN THE ASIAN CONTINENT

Snakebite reports are more common in Asia than in Europe. The incidence of snakebites reported in Asia varies greatly depending on human activity and snake type. According to World Health Organization data (WHO, 2011), nearly 4 million Asian individuals are bitten by snakes yearly. Among these bites, venomous snakes account for about half of the envenomations. Moreover, 50% of the envenomation cases result in the death of the victims (Gutierrez et al., 2017). Among the different regions of the Asian continent, the South Asian, Southeast Asian, and Eastern Asian nations are most affected by snake envenomation (Table 1.1).

1.4.1 South Asia

Snake envenoming is a severe issue in the rural tropical regions of South Asia, including Nepal, Bhutan, Pakistan, Sri Lanka, Bangladesh, and India (WHO, 2011). According to the most conservative estimates, 121,000 envenomings occur yearly in South Asia. The cumulative incidence of snake envenoming in this region ranges from 7.8 to 29.9/100,000, while the mortality rate ranges between 0.9 and 2.2/100,000 (Kasturiratne et al., 2008). However, it is a daunting task to estimate the actual incidence and mortality due to snake envenomation. Because snakebite fatalities are frequently not recorded, hospital records do not accurately reflect the actual rate of snakebite incidence and mortality. For example, a review of the 1999–2003 snakebite data accessible in the hospitals of the Monaragala District of Sri Lanka indicated that, contrary to the hospital records, the absolute number of snakebite deaths around that time was much higher (Fox et al., 2006; Mukherjee, 2021). Similarly, an examination of district hospital data in Nepal revealed that official statistics significantly understated the frequency of snake bites by one order of magnitude (Alirol et al., 2010; Sharma et al., 2003).

In 2000, 15 district hospitals in Nepal recorded 3,189 snakebite cases and 144 cases of snakebite deaths, with an overall death rate of 4.5% (Sharma et al., 2003). Moreover, the latest conservative estimates suggest that around 20,000 envenomings occur annually in Nepal (Kasturiratne et al., 2008). Seventy-eight snakebite cases were reported in Bhutan from June 2013 to July 2015, with a mortality rate of only 0.02% (Koirala et al., 2018). However, in Pakistan, the mortality rate due to snake envenomation is significantly higher at ~20%, with nearly 40,000 envenomation cases reported yearly (Alirol et al., 2010; Kasturiratne et al., 2008). Sri Lanka is also severely affected by snake envenomation; conservative estimates suggest 33,000 envenoming cases per year in Sri Lanka (Kasturiratne et al., 2008). The latest nationwide survey also indicates the occurrence of 27,500 to 36,000 snake envenomations in the island nation, with mortality ranging from 0.2 to 4.4 per 100,000 people (Ediriweera et al., 2016).

Notably, the mortality rate in Bangladesh due to snake envenomation seems quite variable and ranges from 0.5 to 22%. An epidemiological survey conducted by Rahman and co-workers (2010) indicates that the incidence of snakebite among the rural Bangladeshi population is 623.4/100,000 individuals, with an overall mortality rate of ~1% (Rahman et al., 2010). However, earlier reports have suggested significantly higher mortalities of up to 22% with an estimated frequency of 4.3/100,000 persons (Huq et al., 1995; Sarker et al., 1999). A total of 1666 snakebite victims visited the Chittagong Medical College Hospital (CMCH) between 1993 and 2003, and out of them, 28.5% were bitten by venomous snakes, with a relatively lower death rate of 0.5% (Faiz et al., 1995).

Another hospital-based study conducted at Rangpur Medical College and Hospital, Rangpur, Bangladesh, showed that the majority of snake bites (90.09%, n = 100) did not exhibit symptoms of envenomation and that the bulk of venomous bites were accounted for by Cobras (22.23%) and Kraits (77.78%). The study also found that the overall mortality rate was 5.7%, while the same was as high as 54.5% for Cobra and Krait-envenomed patients (Mondal et al., 2012). The latest nationwide cross-sectional epidemiological survey in Bangladesh suggested that an estimated 15,372 (10.98/100,000) individuals were bitten by snakes annually, and among those, 1709 (1.22/100,000) victims succumbed to death with a mortality rate of ~11% (Hossain et al., 2016).

The snakebite scenario in India is very poignant. A thorough epidemiological research on the snakebite problem in India was started by Swaroop and Grab (1954) (Swaroop and Grab, 1954). West Bengal, a province in the eastern zone, had the highest number of snakebite fatality cases, according to their statistical analysis of national data on snakebite from 1940 to 1949. According to Gaitonde and Bhattacharya (1980), Maharashtra, the western zone province of India, has an annual incidence of severe envenomation of almost 70 persons per 100,000 residents and a yearly mortality rate of nearly 2.4 per 10,000 people (Gaitonde and Bhattacharya, 1980). According to Jena and Sarangi (1993), India witnesses over 300,000 cases of snakebite each year, with over 10,000 fatalities (Jena and Sarangi, 1993).

Moreover, as per Chippuax (1998), the number of snakebite victims in India varies between 66 and 163 per 10,000 individuals, with 1.4 to 68 fatalities reported each year. However, most snake bite reports in India are unreliable since they are rarely based on field surveys (Hati et al., 1992). Recent global conservative estimates suggest that India records the highest number of envenomings worldwide at 81,000 cases per year (Kasturiratne et al., 2008). Another epidemiological study on snakebite indicated that approximately 50,000 people die from snakebite each year in India (Mohapatra et al., 2011).

Most snakebite cases (97%) occur near rural agricultural fields. The latest epidemiological survey suggests India witnessed 1.2 million snakebite deaths (about 58,000 per year) between 2000 and 2019 (Suraweera et al., 2020). The snakebite death rates in different Indian states from 2001 to 2014 are summarized in Table 1.2. Notably, the overall age-standardized rates of snakebite envenoming deaths in India have decreased to 4.0 per 100,000 (2.3–5.0) in 2019 from 7.3 per 100,000 (4.1–8.8) in 1990 (GBD, 2022).

1.4.2 Southeast Asia

With 111,000 envenomation cases annually, Southeast Asia is another hotspot for snakebite incidence and mortality. The cumulative incidence of snake envenoming in Southeast Asia ranges from 18.8 to 84.6/100,000, while the mortality rate ranges between 0.1 and 3.2/100,000 (Kasturiratne et al., 2008). However, the most recent epidemiological report from the ASEAN (Association of Southeast Asian Nations) countries, including Vietnam, Myanmar, Thailand, Philippines, Malaysia, Indonesia, and Lao PDR estimated nearly 242,648 (209,810–291,023) envenomation cases and 15,909 (7,592–33,949) deaths, with a yearly prevalence that ranged from 32.89 to 45.62 per 100,000 person per year in Southeast Asia (Patikorn et al., 2022).

The estimated incidence of snakebite was recorded as the highest in Lao PDR (200.00 persons per 100,000 people). Malaysia shows the minimum number of snakebites, where 10.68 persons per 100,000 population were bitten by snakes (Figure 1.2). More significantly, it was discovered that 69% of victims (n = 161,835) were not treated with antivenom. It is striking to note that only 4.2% of the antivenom-indicated victims in Lao PDR were treated with antivenom. These victims sought treatment either from traditional healers or from medical facilities where there was a shortage of antivenom supply and, therefore, were not treated with antivenom. On the contrary, antivenom was administered in 94.2% and 99.9% of the antivenom-indicated victims in Malaysia and Thailand, respectively. Concurrently, the mortality rate was also the lowest in Malaysia and Thailand (0.006 per 100,000 population), while it was several folds higher in Lao PDR (14.04 per 100,000 population) (Figure 1.2) (Patikorn et al., 2022).

1.4.3 Eastern Asia

The burden of snakebite in Eastern Asia is relatively lower compared to South and Southeast Asian regions. The cumulative incidence of snake envenoming in Eastern

TABLE 1.2
Data on Snakebite Death Rates in Various Indian States From 2001 to 2014

Indian States	Snakebite Deaths in Million Death Study	Annual Average Standardized Death Rate/100,000			Estimated Deaths for 2001–2014
		2001–2004	2005–2009	2010–2014	
Higher burden states	1726	5.9	6.1	6.2	557.4
Andhra Pradesh	271	8.5	7.3	5.6	82.9
Bihar	321	5.6	7.6	8.9	101.9
Odisha	191	7.5	7.2	5.9	40.3
Madhya Pradesh	195	6.7	7.7	6.0	67.8
Uttar Pradesh	322	5.2	5.9	6.0	153.6
Rajasthan	192	4.9	6.7	5.0	52.1
Gujarat	176	4.1	4.8	5.1	38.8
Jharkhand	58	4.9	2.0	7.1	20.1
Lower burden states	1107	3.7	3.1	2.1	249.9
Chhattisgarh	42	6.0	6.5	2.5	16.8
Jammu & Kashmir	64	5.3	7.0	0.9	7.0
Tamil Nadu	176	6.1	3.4	3.0	42.1
Karnataka	137	5.6	3.3	2.9	33.0
Maharashtra	147	4.2	3.7	2.6	56.0
West Bengal	188	4.1	3.3	2.9	42.7
Punjab	67	2.9	3.1	4.0	14.5
Haryana	45	2.9	3.3	1.8	9.5
Assam	27	2.8	0.7	2.1	7.3
Northeastern states	37	2.3	0.9	0.7	2.4
Kerala	43	1.8	1.3	0.5	6.5
All other states	134	4.3	3.9	3.2	12.2
All India	**2833**	**5.1**	**4.9**	**4.5**	**807.5**

Source: Adopted from Suraweera et al. (2020).

Asia ranges from 0.3 to 15.7/100,000, while the mortality rate ranges between 0.03 and 0.3/100,000 (Kasturiratne et al., 2008). Among the different East Asian nations, the incidence of snake envenomation is the highest for Taiwan, at 13.6 to 26.6 per 100,000 people. A more recent nationwide epidemiological study by Chen and co-workers (2015) suggested that the annual incidence of snake envenomation in Taiwan may be even higher at 40.5 per 100,000 residents (Chen et al., 2015). Moreover, the mortality rate in Taiwan is also the highest (0.03 to 2.5/100,000 population) among the East Asian countries (Kasturiratne et al., 2008).

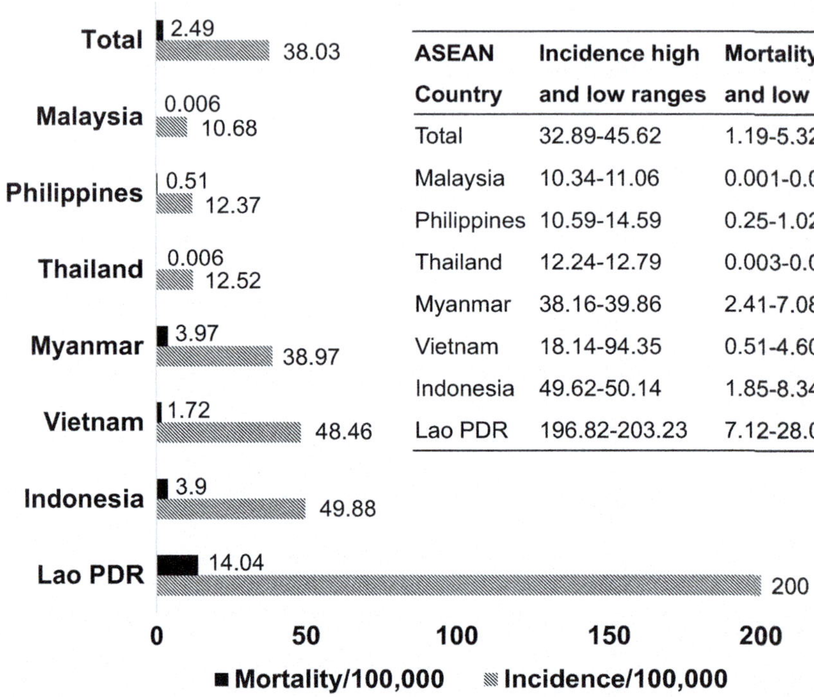

ASEAN Country	Incidence high and low ranges	Mortality high and low ranges
Total	32.89-45.62	1.19-5.32
Malaysia	10.34-11.06	0.001-0.019
Philippines	10.59-14.59	0.25-1.02
Thailand	12.24-12.79	0.003-0.009
Myanmar	38.16-39.86	2.41-7.08
Vietnam	18.14-94.35	0.51-4.60
Indonesia	49.62-50.14	1.85-8.34
Lao PDR	196.82-203.23	7.12-28.03

FIGURE 1.2 Estimated annual incidence and mortality due to snakebite in ASEAN countries. **(Epidemiological data retrieved from Patikorn et al., 2022.)**

1.5 COBRA ENVENOMATION IN ASIA – KEY ISSUES

Over 150 venomous snakes from the Viperidae, Elapidae, and Colubridae families inhabit the Asian continent. Among the myriad venomous snake species, Cobras belonging to the Elapidae family of snakes are responsible for many envenomation cases, mortalities, and morbidities in Asia. Eleven species of Cobras (*Naja*) are found in Asia, and all except *N. sagittifera* are classified as Category I medically necessary snakes by the WHO (see Chapter 4 for details). These data signify the severity and importance of Cobra envenomation in Asian countries.

Although discrete epidemiological data on Cobra envenomation is scarce, few studies have demonstrated that envenomation by a Cobra species is quite frequent in some Asian nations. For instance, out of the 1006 envenomed victims included in a retrospective study over two years in Thailand, Cobra envenomation (n = 271) was found to be the most common type of snakebite in victims with neurotoxic symptoms (n = 356). Furthermore, Cobra was the second-most frequent culprit (~27%) involved in envenomation, preceded by green pit vipers (~33%) (Sriapha et al., 2022). Similarly, 30% of snake envenomation cases in Taiwan usually result from Chinese Cobra (*N. atra*) encounters. Notably, although the most common cases in Taiwan are caused by crotalid snake (*Trimeresurus stejnegeri* and *Protobothrops mucrosquamatus*) envenomations, those by the Chinese Cobra cause more systemic severe effects as well as wound complications (Mao et al., 2018).

In a retrospective study of 30 snake-envenomed victims in western India, Cobra's (*N. naja*) envenomation frequency was nearly 24%. However, it is alarming to note that two out of the seven patients died upon arrival at the hospital; another patient suffered cardiac arrest and succumbed to death (Bawaskar and Bawaskar, 2004). Such episodes underscore the need for a prompt medical response toward Cobra-envenomed victims. In a recent epidemiological study from North-east India that involved 139 snake-envenomed patients, the frequency of Cobra (*N. kaouthia*) envenomation was reported to be ~16%; however, no mortality was recorded (Kakati et al., 2023). Similarly, out of 884 snake-envenomed patients, 70 (~8%) were identified as victims of bites by *N. kaouthia*, and two cases had fatal outcomes (Faiz et al., 2017).

Most Cobras inhabit areas close to human dwellings, such as agricultural fields and watercourses, where they have easy access to prey. It has been noted across several epidemiological studies that the dry months of November through January witness the lowest frequency of Cobra bites, while when temperatures and humidity elevate to their peak during the monsoon season of April through August, the frequency of Cobra envenomation increases significantly (Faiz et al., 2017; Kakati et al., 2023; Mao et al., 2018; Sriapha et al., 2022). The reason is primarily because snakes aggressively search for food and mate during the monsoon season, which is their breeding season (Musah et al., 2019). At the same time, there are also a lot of intensive agricultural activities during this time of the year, which consequently increases the frequency of snake-human encounters and the number of snakebite cases.

Agricultural activity is one of Asia's primary sources of revenue and is thus necessary. However, barefoot farming should be avoided because the lower limbs are most susceptible to snakebite. In addition, taking necessary precautions, such as wearing protective gear and gumboots during the rainy season, can prevent snake encounters. Notably, the majority of envenomed victims are males between the ages of 18 and 60 who work in agriculture. This is a severe socioeconomic issue because many of the victims may be the sole earners in their families. Epidemiological studies also show that most Cobra species are diurnal, meaning most bites occur during the day. However, *N. kaouthia* is a nocturnal species, and envenomation of this Cobra species has been documented primarily during the early morning and late evening (Faiz et al., 2017).

1.6 CONCLUSION

To summarize, the Asian continent has the highest rates of snakebite incidence and fatality, most likely due to a lack of systematic protection measures at the national and public levels. Furthermore, among Asian countries and around the world, India has the highest incidence and mortality rate from snakebite. Unfortunately, most snakebite epidemiology data comes from hospital records and is dispersed across the continent. Furthermore, many snakebite patients die during their transfer to the hospital, or the victims rely on traditional healers for treatment. Thus, due to a lack of systematic research, it is impossible to make a definite judgment on the severity of envenomation in Asian countries where snakebite is a serious concern. It is also worth noting that Cobra bites have a significant impact on the Asian continent, accounting for a large number of envenomations, deaths, and morbidities.

REFERENCES

Akani, G., Luiselli, L., Angelici, F., Corti, C., Zuffi, M.A.L., 2001. The case of rainforest stiletto snakes (genus *Atractaspis*) in southern Nigeria. Evidence of diverging foraging strategies in grossly sympatric snakes with homogeneous body architecture? *Ethology Ecology and Evolution* 13(1), 89–94.

Ali, S.A., Yang, D.C., Jackson, T.N., Undheim, E.A., Koludarov, I., Wood, K., Jones, A., Hodgson, W.C., McCarthy, S., Ruder, T., Fry, B.G., 2013. Venom proteomic characterization and relative antivenom neutralization of two medically important Pakistani elapid snakes (*Bungarus sindanus* and *Naja naja*). *Journal of Proteomics* 89, 15–23.

Alirol, E., Sharma, S.K., Bawaskar, H.S., Kuch, U., Chappuis, F., 2010. Snake bite in South Asia: A review. *PLoS Neglected Tropical Diseases*, 4, e603.

Bawaskar, H.S., Bawaskar, P.H., 2004. Envenoming by the common krait (Bungarus caeruleus) and Asian cobra (*Naja naja*): Clinical manifestations and their management in a rural setting. *Wilderness & Environmental Medicine*, 15, 257–266.

Casewell, N.R., Wuster, W., Vonk, F.J., Harrison, R.A., Fry, B.G., 2013. Complex cocktails: The evolutionary novelty of venoms. *Trends in Ecology & Evolution*, 28, 219–229.

Chanda, A., Kalita, B., Patra, A., Senevirathne, W., Mukherjee, A.K., 2019. Proteomic analysis and antivenomics study of Western India *Naja naja* venom: Correlation between venom composition and clinical manifestations of cobra bite in this region. *Expert Review of Proteomics*, 16, 171–184.

Chanda, A., Mukherjee, A.K., 2020. Quantitative proteomics to reveal the composition of Southern India spectacled cobra (*Naja naja*) venom and its immunological cross-reactivity towards commercial antivenom. *International Journal of Biological Macromolecules*, 160, 224–232.

Chanda, A., Patra, A., Kalita, B., Mukherjee, A.K., 2018. Proteomics analysis to compare the venom composition between *Naja naja* and *Naja kaouthia* from the same geographical location of eastern India: Correlation with pathophysiology of envenomation and immunological cross-reactivity towards commercial polyantivenom. *Expert Review of Proteomics*, 15, 949–961.

Chen, C.-K., Lin, C.-C., Shih, F.-Y., Chaou, C.-H., Lin, J.C.-C., Lai, T.-I., Tseng, C.-Y., Fang, C.-C., 2015. Population-based study of venomous snakebite in Taiwan. *Journal of Acute Medicine*, 5, 38–42.

Cheng, A.C., Currie, B.J., 2004. Venomous snakebites worldwide with a focus on the Australia-Pacific region: Current management and controversies. *Journal of Intensive Care Medicine*, 19, 259–269.

Chippaux, J.P., 1998. Snake-bites: Appraisal of the global situation. *Bulletin of the World Health Organization*, 76, 515–524.

Chippaux, J.P., 2012. Epidemiology of snakebites in Europe: A systematic review of the literature. *Toxicon*, 59, 86–99.

Chippaux, J.P., Diouf, A., Massougbodji, A., Stock, R.P., Kane, O., Dieye, A.M., Lam, F.A., Mbaye, S.M., Parra, H.J., 2011. The 4th international conference on snake bites and scorpion stings in Dakar, April 25–29, 2011. *Medecine Tropicale : Revue du Corps de Sante Colonial* 71, 431–433.

Da Silva, C., Jorge, M., Ribeiro, L., 2003. Epidemiology of snakebite in a central region of Brazil. *Toxicon* 41, 251–255.

Daltry, J.C., Wuster, W., Thorpe, R.S., 1996. Diet and snake venom evolution. *Nature* 379, 537–540.

de Haro, L., 2012. Management of snakebites in France. *Toxicon* 60, 712–718.

Dutta, S., Chanda, A., Kalita, B., Islam, T., Patra, A., Mukherjee, A.K., 2017. Proteomic analysis to unravel the complex venom proteome of eastern India *Naja naja*: Correlation of venom composition with its biochemical and pharmacological properties. *Journal of Proteomics* 156, 29–39.

Ediriweera, D.S., Kasturiratne, A., Pathmeswaran, A., Gunawardena, N.K., Wijayawickrama, B.A., Jayamanne, S.F., Isbister, G.K., Dawson, A., Giorgi, E., Diggle, P.J., Lalloo, D.G., de Silva, H.J., 2016. Mapping the Risk of Snakebite in Sri Lanka - A National Survey with Geospatial Analysis. *PLoS Neglected Tropical Diseases* 10, e0004813.

Faiz, M., Rahman, M., Hussain, A., Yunus, E., Das, J., Karim, S., Khan, I., Mollah, A., Huq, J., Chowdhury, M., 1995. A hospital based study of snake bite in Chittagong Medical College. *Journal of Bangladesh College of Physicians & Surgeons* 13, 3–8.

Faiz, M.A., Ahsan, M.F., Ghose, A., Rahman, M.R., Amin, R., Hossain, M., Tareq, M.N.U., Jalil, M.A., Kuch, U., Theakston, R.D.G., Warrell, D.A., Harris, J.B., 2017. Bites by the Monocled Cobra, *Naja kaouthia*, in Chittagong Division, Bangladesh: Epidemiology, Clinical Features of Envenoming and Management of 70 Identified Cases. *The American Journal of Tropical Medicine and Hygiene* 96, 876–884.

Fox, J.W., Serrano, S.M., 2008. Exploring snake venom proteomes: Multifaceted analyses for complex toxin mixtures. *Proteomics* 8, 909–920.

Fox, S., Rathuwithana, A.C., Kasturiratne, A., Lalloo, D.G., de Silva, H.J., 2006. Underestimation of snakebite mortality by hospital statistics in the Monaragala District of Sri Lanka. *Transactions of the Royal Society of Tropical Medicine and Hygiene* 100, 693–695.

Fry, B.G., 2005. From genome to "venome": Molecular origin and evolution of the snake venom proteome inferred from phylogenetic analysis of toxin sequences and related body proteins. *Genome Research* 15, 403–420.

Gaitonde, B., Bhattacharya, S., 1980. An epidemiological survey of snake bite cases in India. *Snake* 12, 129–133.

GBD, 2022. Global mortality of snakebite envenoming between 1990 and 2019. *Nature Communications* 13, 6160.

Gutierrez, J.M., Calvete, J.J., Habib, A.G., Harrison, R.A., Williams, D.J., Warrell, D.A., 2017. Snakebite envenoming. *Nature reviews. Disease Primers* 3, 17079.

Gutierrez, J.M., Theakston, R.D., Warrell, D.A., 2006. Confronting the neglected problem of snake bite envenoming: The need for a global partnership. *PLoS Medicine* 3, e150.

Gutierrez, J.M., Williams, D., Fan, H.W., Warrell, D.A., 2010. Snakebite envenoming from a global perspective: Towards an integrated approach. *Toxicon: Official Journal of the International Society on Toxinology* 56, 1223–1235.

Habib, A.G., Kuznik, A., Hamza, M., Abdullahi, M.I., Chedi, B.A., Chippaux, J.P., Warrell, D.A., 2015. Snakebite is Under Appreciated: Appraisal of Burden from West Africa. *PLoS Neglected Tropical Diseases* 9, e0004088.

Hati, A.K., Mandal, M., De, M.K., Mukherjee, H., Hati, R.N., 1992. Epidemiology of snake bite in the district of Burdwan, West Bengal. *Journal of the Indian Medical Association* 90, 145–147.

Hill, R.E., Mackessy, S.P., 2000. Characterization of venom (Duvernoy's secretion) from twelve species of colubrid snakes and partial sequence of four venom proteins. *Toxicon* 38, 1663–1687.

Hossain, J., Biswas, A., Rahman, F., Mashreky, S.R., Dalal, K., Rahman, A., 2016. Snakebite Epidemiology in Bangladesh: A national community based health and injury survey. *Health* 8, 479–486.

Huq, F., Islam, M., Sarker, M., Chowdhury, B., Ali, M., Kabir, M., 1995. Epidemiology of snake bite in Bangladesh. *Bangladesh Journal of Zoology* 23, 61–64.

Jena, I., Sarangi, A., 1993. *Snakes of medical importance and snake-bite treatment*. Ashish Publication.

Junqueira-de-Azevedo, I.L., Bastos, C.M., Ho, P.L., Luna, M.S., Yamanouye, N., Casewell, N.R., 2015. Venom-related transcripts from *Bothrops jararaca* tissues provide novel molecular insights into the production and evolution of snake venom. *Molecular Biology and Evolution* 32, 754–766.

Kakati, H., Giri, S., Patra, A., Taye, S.J., Agarwalla, D., Boruah, H., Choudhary, G., Kalita, B., Mukherjee, A. K., 2023. A retrospective analysis of epidemiology, clinical features of envenomation, and in-patient management of snakebites in a model secondary hospital of Assam, *North-East India. Toxicon* 230, 107175.

Kalita, B., Patra, A., Das, A., Mukherjee, A.K., 2018. Proteomic Analysis and Immuno-Profiling of Eastern India Russell's Viper (*Daboia russelii*) Venom: Correlation between RVV Composition and Clinical Manifestations Post RV Bite. *Journal of Proteome Research* 17, 2819–2833.

Kalita, B., Patra, A., Mukherjee, A.K., 2017. Unraveling the proteome composition and immuno-profiling of Western India Russell's viper venom for in-depth understanding of its pharmacological properties, clinical manifestations, and effective antivenom treatment. *Journal of Proteome Research* 16, 583–598.

Kalita, B., Saviola, A.J., Mukherjee, A.K., 2021. From venom to drugs: A review and critical analysis of Indian snake venom toxins envisaged as anticancer drug prototypes. Drug Discovery Today 26, 993–1005.

Kasturiratne, A., Wickremasinghe, A.R., de Silva, N., Gunawardena, N.K., Pathmeswaran, A., Premaratna, R., Savioli, L., Lalloo, D.G., de Silva, H.J., 2008. The global burden of snakebite: A literature analysis and modelling based on regional estimates of envenoming and deaths. *PLoS Medicine* 5, e218.

Koh, C.Y., Kini, R.M., 2012. From snake venom toxins to therapeutics-cardiovascular examples. *Toxicon* 59, 497–506.

Koirala, B.K., Koirala, J., Sapkota, S., 2018. Retrospective study on epidemiology of snakebites in Sarpang District, southern Bhutan. *Journal of Threatened Taxa* 10, 12749–12754.

Lawson, R., Slowinski, J.B., Crother, B.I., Burbrink, F.T., 2005. Phylogeny of the Colubroidea (Serpentes): New evidence from mitochondrial and nuclear genes. *Molecular Phylogenetics and Evolution* 37, 581–601.

Mackessy, S.P., 2002. Biochemistry and pharmacology of colubrid snake venoms. *Journal of Toxicology - Toxin Reviews* 21, 43–83.

Mackessy, S.P., 2010. Evolutionary trends in venom composition in the western rattlesnakes (*Crotalus viridis* sensu lato): Toxicity vs. tenderizers. *Toxicon* 55, 1463–1474.

Mackessy, S.P., Baxter, L.M., 2006. Bioweapons synthesis and storage: The venom gland of front-fanged snakes. *Zoologischer Anzeiger-A Journal of Comparative Zoology* 245, 147–159.

Mackessy, S.P., Saviola, A.J., 2016. *Understanding biological roles of venoms among the Caenophidia: The importance of rear-fanged snakes*. Integrative and Comparative Biology 56, 1004-1021.

Mao, Y.C., Liu, P.Y., Chiang, L.C., Lai, C.S., Lai, K.L., Ho, C.H., Wang, T.H., Yang, C.C., 2018. Naja atra snakebite in Taiwan. *Clinical Toxicology* 56, 273–280.

Mathers, C.D., Ezzati, M., Lopez, A.D., 2007. Measuring the burden of neglected tropical diseases: The global burden of disease framework. *PLoS Neglected Tropical Diseases* 1, e114.

McDowell, S.B., 1972. The evolution of the tongue of snakes, and its bearing on snake origins. *Evolutionary Biology* 6, 191–273.

Mebs, D., 2002. *Scorpions and snakes, such as cobras, mambas and vipers made the African continent famous for venomous animals. Bulletin de la Societe de Pathologie Exotique* 95, 131.

Mohapatra, B., Warrell, D.A., Suraweera, W., Bhatia, P., Dhingra, N., Jotkar, R.M., Rodriguez, P.S., Mishra, K., Whitaker, R., Jha, P., Million Death Study, C., 2011. Snakebite mortality in India: A nationally representative mortality survey. *PLoS Neglected Tropical Diseases* 5, e1018.

Mondal, R.N., Chowdhury, F.R., Rani, M., Mohammad, N., Monjurul Islam, M., Ashraful Haque, M., Abul Faiz, M., 2012. Pre-hospital and hospital management practices and circumstances behind venomous snakebite in northwestern part of Bangladesh. *Asia Pacific Journal of Medical Toxicology* 1, 18–21.

Mukherjee, A., Ghosal, S., Maity, C., 2000. Some biochemical properties of Russell's viper (*Daboia russelli*) venom from Eastern India: Correlation with clinico-pathological manifestation in Russell's viper bite. *Toxicon* 38, 163–175.

Mukherjee, A.K., 2014. The pro-coagulant fibrinogenolytic serine protease isoenzymes purified from Daboia russelii russelii venom coagulate the blood through factor V activation: Role of glycosylation on enzymatic activity. *PLoS One* 9, e86823.

Mukherjee, A.K., 2021. *Prevention and treatment of the "Big Four" snakebite in India, the 'big Four' snakes of India: Venom composition, pharmacological properties and treatment of envenomation*. Springer, pp. 145–161.

Mukherjee, A.K., Kalita, B., Mackessy, S.P., 2016. A proteomic analysis of Pakistan *Daboia russelii russelii* venom and assessment of potency of Indian polyvalent and monovalent antivenom. *Journal of Proteomics* 144, 73–86.

Mukherjee, A.K., Mackessy, S.P., 2013. Biochemical and pharmacological properties of a new thrombin-like serine protease (Russelobin) from the venom of Russell's Viper (*Daboia russelii russelii*) and assessment of its therapeutic potential. *Biochimica et Biophysica Acta* 1830, 3476–3488.

Mukherjee, A.K., Mackessy, S.P., 2021. Prevention and improvement of clinical management of snakebite in Southern Asian countries: A proposed road map. *Toxicon* 200, 140–152.

Mukherjee, A.K., Maity, C.R., 2002. Biochemical composition, lethality and pathophysiology of venom from two cobras - *Naja naja* and *N. kaouthia*. *Comparative Biochemistry and Physiology. Part B, Biochemistry & Molecular Biology* 131, 125–132.

Munekiyo, S.M., Mackessy, S.P., 1998. Effects of temperature and storage conditions on the electrophoretic, toxic and enzymatic stability of venom components. *Comparative Biochemistry and Physiology. Part B, Biochemistry & Molecular Biology* 119, 119–127.

Musah, Y., Ameade, E.P.K., Attuquayefio, D.K., Holbech, L.H., 2019. Epidemiology, ecology and human perceptions of snakebites in a savanna community of northern Ghana. *PLoS Neglected Tropical Diseases* 13, e0007221.

Ohno, M., Menez, R., Ogawa, T., Danse, J., Shimohigashi, Y., Fromen, C., Ducancel, F., Zinn-Justin, S., Boulain, J.-C., Tamiya, T., 1997. Molecular evolution of snake toxins: Is the functional diversify of snake toxins associated with a mechanism of accelerated evolution? *Progress in Nucleic Acid Research and Molecular Biology* 59, 307–364.

Patikorn, C., Blessmann, J., Nwe, M.T., Tiglao, P.J.G., Vasaruchapong, T., Maharani, T., Doan, U.V., Zainal Abidin, S.A., Ismail, A.K., Othman, I., Taychakhoonavudh, S., Chaiyakunapruk, N., 2022. Estimating economic and disease burden of snakebite in ASEAN countries using a decision analytic model. *PLoS Neglected Tropical Diseases* 16, e0010775.

Patra, A., Kalita, B., Chanda, A., Mukherjee, A.K., 2017. Proteomics and antivenomics of *Echis carinatus carinatus* venom: Correlation with pharmacological properties and pathophysiology of envenomation. *Scientific Reports* 7, 17119.

Peichoto, M.E., Tavares, F.L., Santoro, M.L., Mackessy, S.P., 2012. Venom proteomes of South and North American opisthoglyphous (Colubridae and Dipsadidae) snake species: A preliminary approach to understanding their biological roles. *Comparative Biochemistry and Physiology. Part D, Genomics & Proteomics* 7, 361–369.

Rahman, R., Faiz, M.A., Selim, S., Rahman, B., Basher, A., Jones, A., d'Este, C., Hossain, M., Islam, Z., Ahmed, H., Milton, A.H., 2010. Annual incidence of snake bite in rural bangladesh. *PLoS Neglected Tropical Diseases* 4, e860.

Sarker, M., Sarker, N., Patwary, S., 1999. Epidemiological survey of snakebite incidences in Bangladesh. *Dhaka University Journal of Biological Sciences* 8, 53–68.

Sasa, M., Vazquez, S., 2003. Snakebite envenomation in Costa Rica: A revision of incidence in the decade 1990–2000. *Toxicon* 41, 19–22.

Sharma, S.K., Khanal, B., Pokhrel, P., Khan, A., Koirala, S., 2003. Snakebite-reappraisal of the situation in Eastern Nepal. *Toxicon* 41, 285–289.

Shine, R., Branch, W.R., Harlow, P.S., Webb, J.K., Shine, T., 2006. Biology of burrowing asps (Atractaspididae) from Southern Africa. *Copeia* 2006, 103–115.
Smith, J.R., Yeo, W.C., 1977. Automatic classification of the cats vigilance state. *Computers in Biology and Medicine* 7, 199–207.
Sriapha, C., Rittilert, P., Vasaruchapong, T., Srisuma, S., Wananukul, W., Trakulsrichai, S., 2022. Early adverse reactions to snake antivenom: Poison center data analysis. *Toxins* 14, 694.
Stahel, E., Wellauer, R., Freyvogel, T.A., 1985. Poisoning by domestic vipers (Vipera berus and Vipera aspis). A retrospective study of 113 patients. *Schweizerische Medizinische Wochenschrift* 115, 890–896.
Stanley, S.M., 1979. *Macroevolution, pattern and process*. Editor, W. H. Freeman.
Stewart, C.J., 2003. Snake bite in Australia: First aid and envenomation management. *Accident and Emergency Nursing* 11, 106–111.
Suraweera, W., Warrell, D., Whitaker, R., Menon, G., Rodrigues, R., Fu, S.H., Begum, R., Sati, P., Piyasena, K., Bhatia, M., Brown, P., Jha, P., 2020. Trends in snakebite deaths in India from 2000 to 2019 in a nationally representative mortality study. *eLife* 9, e54076.
Swaroop, S., Grab, B., 1954. Snakebite mortality in the world. *Bulletin of the World Health Organization* 10, 35.
Tan, C.H., Tan, N.H., Sim, S.M., Fung, S.Y., Gnanathasan, C.A., 2015. Proteomic investigation of Sri Lankan hump-nosed pit viper (*Hypnale hypnale*) venom. *Toxicon* 93, 164–170.
Thakur, R., Mukherjee, A.K., 2017. Pathophysiological significance and therapeutic applications of snake venom protease inhibitors. *Toxicon* 131, 37–47.
Viala, V.L., Hildebrand, D., Trusch, M., Fucase, T.M., Sciani, J.M., Pimenta, D.C., Arni, R.K., Schluter, H., Betzel, C., Mirtschin, P., Dunstan, N., Spencer, P.J., 2015. Venomics of the Australian eastern brown snake (*Pseudonaja textilis*): Detection of new venom proteins and splicing variants. *Toxicon* 107, 252–265.
Warrell, D., 1996. Clinical features of envenoming from snakebites. *Toxicon* 2, 144.
Warrell, D., 1999. WHO/SEARO Guidelines for the clinical management of snake bites in the Southeast Asian region. *The Southeast Asian Journal of Tropical Medicine and Public Health* 30(Suppl 1), 1–85.
Warrell, D.A., 2013. Redi award lecture: Clinical studies of snake-bite in four tropical continents. *Toxicon* 69, 3–13.
Weldon, C.L., Mackessy, S.P., 2010. Biological and proteomic analysis of venom from the Puerto Rican Racer (Alsophis portoricensis: Dipsadidae). *Toxicon* 55, 558–569.
White, J., 2005. Snake venoms and coagulopathy. *Toxicon* 45, 951–967.
WHO, 2007. *Rabies and envenomings: A neglected public health issue*. World Health Organization Geneva.
WHO, 2011. Neglected tropical diseases: Snakebite. http://www.who.int/neglected_diseases/diseases/snakebites/en/201
Wong, K.Y., Tan, C.H., Tan, K.Y., Quraishi, N.H., Tan, N.H., 2018. Elucidating the biogeographical variation of the venom of *Naja naja* (spectacled cobra) from Pakistan through a venom-decomplexing proteomic study. *Journal of Proteomics* 175, 156–173.

2 Evolution and Conservation of Cobras

2.1 INTRODUCTION

Cobras, belonging to the family Elapidae, are venomous snakes known for their iconic hooded appearance and the ability to deliver a potent neurotoxic venom. These snakes play a crucial role in the ecosystem as both predators and prey, contributing to the delicate balance of nature. Their importance lies in their ecological significance and cultural and religious symbolism, particularly in regions where Cobras are native. In many cultures, Cobras are revered and feared, often associated with deities or mythical creatures, making them an integral part of folklore and traditions (Mukherjee, 2021a).

The evolution of Cobras, one of the most venomous snake species in the world, is a fascinating subject. These serpents have undergone significant changes over millions of years to adapt to their environments and develop impressive defensive mechanisms. From their African origins, Cobras have spread across continents, diversifying into various species with unique physical characteristics and venom compositions (Mukherjee, 2021b). Understanding the evolutionary journey of Cobras can shed light on their remarkable survival strategies and the intricate relationship between predators and prey in the animal kingdom. Moreover, studying Cobras and their evolutionary history can provide valuable insights into the mechanisms behind venom evolution, offering potential applications in medical research and developing antivenom treatments.

The following sub-sections of this chapter explore the ecological significance of Cobras, particularly in Asia. Additionally, this chapter explains the evolutionary history of Cobras and the mechanisms behind venom evolution. This research aims to gain valuable insights that can be applied in medical research and developing antivenom treatments. By understanding the importance of Cobras (Mukherjee, 2021a) in various aspects, we can enhance our knowledge and contribute to the conservation and protection of these fascinating creatures.

2.2 EVOLUTIONARY HISTORY OF COBRAS

Fossil records suggest that the ancestors of modern Cobras first appeared in the late Cretaceous period. These early Cobras were small and lacked the iconic hood characteristic of their descendants. Over time, as they adapted to different environments, Cobras must have undergone significant morphological changes, including

DOI: 10.1201/9781003598121-2

developing their hood and venom apparatus (Kardong et al., 2009; Logan et al., 2021). It is believed that these adaptations played a crucial role in their survival and success as predators. Studying Cobras' origin and early evolution sheds light on their fascinating history. It provides valuable insights into the mechanisms behind venom evolution, potentially leading to advancements in medical research and the development of more effective antivenom treatments.

Previously, all Asian *Naja* species were categorized under *Naja naja* (Wüster, 1996). However, over the past two decades, taxonomic revisions and species discoveries within Asian and African *Naja* species have recognized approximately 30 species today (Ceriaco et al., 2017; Wallach et al., 2009). The genus *Naja* is now subdivided into four subgenera: Uraeus (African non-spitting), Boulengerina (African forest), Afronaja (African spitting), and Naja (Asian spitting and non-spitting) (Figure 2.1) (Wallach et al., 2009; Wüster et al., 2007).

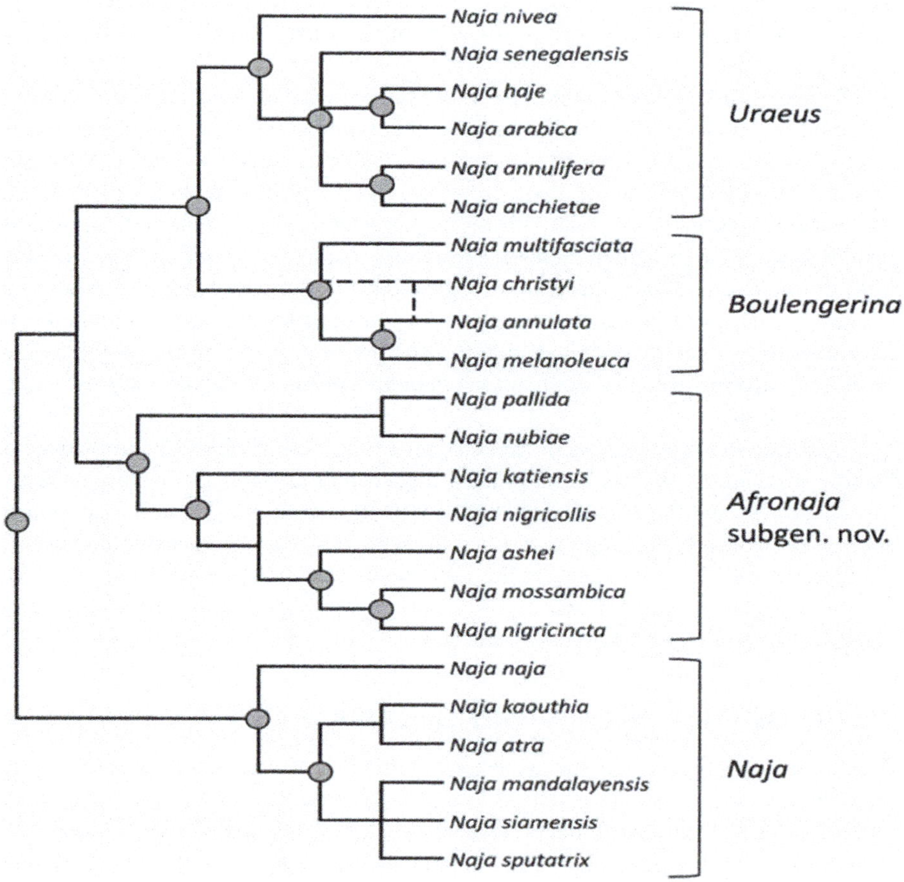

FIGURE 2.1 Schematic representation of phylogeny of the four sub-genera of the genus *Naja*. Grey circles indicate nodes with ≥ 95% Bayesian posterior probability support.

(Redrawn from Wüster et al., 2007.)

2.3 ADAPTATION AND DIVERSIFICATION OF COBRAS OVER TIME

Exploring Cobra's adaptation and diversification reveals how these snakes have evolved to thrive in various habitats and ecological niches. As Cobras spread across different regions, they encountered unique challenges and opportunities, leading to further diversification and specialization. For instance, some Cobras from a particular geographical location developed more potent venom than others to capture larger prey available in that region (Mukherjee, 2020; Mukherjee, 2021a; Senji Laxme et al., 2021). The exploration of Cobras' adaptation and diversification showcases the incredible diversity of these snakes and highlights the intricate interplay between their biology and their environments.

The study on niche evolution (conservatism or divergence) in lineage diversification has revealed that the Chinese Cobra *(Naja atra)* has undergone divergence into three distinct lineages: Lineage E in eastern China, Lineage S in southern China and Vietnam, and Lineage W in western China. It was revealed that the three lineages of *N. atra* occupied the same ecological niche space between Lineages E and S/W. The climatic niches of Lineages S and W were found to be a specialized subset of the climatic niche of Lineage E (Zhu et al., 2022). The authors hypothesized that the disparity in ecological niches between Lineages S and W resulted from geographical barriers that restricted the exchange of genetic material. The researchers presented evidence supporting the idea that lineage diversification is influenced by geographical separation and climatic niches' emergence. They propose that the initial diversity in niches between Lineages S and W, followed by a tendency to maintain them, leads to divergence among different lineages (Zhu et al., 2022).

2.4 SIGNIFICANT EVOLUTIONARY MILESTONES IN THE LINEAGE OF COBRAS

One significant evolutionary milestone in the lineage of Cobras is the development of their iconic hood (Jones, 2017; Kardong, 1982; Young and Kardong, 2010). This unique adaptation allows Cobras to display a threatening posture by spreading the skin of their necks, making them appear larger and more intimidating to potential predators or rivals. The evolution of this hood has likely provided Cobras with a competitive advantage in territorial disputes and during courtship rituals.

Another significant milestone in Cobra's evolution is the refinement of their venom delivery system (Jackson, 2003; Kardong, 1982). Over time, Cobras have developed specialized fangs and venom glands that deliver potent toxins with great precision and efficiency (see Chapter 3 for more details). This adaptation has enabled Cobras to become formidable predators capable of swiftly subduing and killing their prey.

2.5 MORPHOLOGICAL AND PHYSIOLOGICAL EVOLUTION

2.5.1 Evolution of Physical Characteristics and Anatomical Features in Cobras: Enhancing Survivability and Hunting Abilities

Cobras have undergone significant morphological changes, such as elongated bodies, flattened heads, hoods, and different body colours that can expand to intimidate predators or competitors and adapt them to a particular environment (Panagides et al., 2017). These adaptations help Cobras in various ways, increasing their striking range and making them appear larger and more threatening (Jones, 2017).

Moreover, Cobras have also developed specialized scales on their bellies, which aid in locomotion and allow them to move swiftly across different terrains, such as sandy dunes or dense vegetation. These adaptations have contributed to the widespread distribution of Cobras across various habitats, making them highly successful predators in their respective ecosystems (Marais, 2022). For example, when a Cobra encounters a small mammal like a rat, its elongated body and flattened head allow it to manoeuvre swiftly through narrow spaces and ambush its prey. Its expanded hood intimidates the rat, making it freeze in fear or divert its attention from the Cobra's striking range. Additionally, Cobras have evolved hooded structures on their necks that can be spread out when threatened, making them appear larger and more intimidating to potential predators or rivals (Young and Kardong, 2010). This defensive mechanism increases their chances of survival in territorial disputes.

2.5.2 The Physiological Adaptations That Have Allowed Cobras to Thrive in Different Environments

Physiologically, Cobras have developed a higher resistance to their venom, allowing them to handle and inject it without harming themselves. Their venom composition has also evolved in different geographical regions to target specific prey, making them more efficient hunters depending on the prey available in that region (Chanda et al., 2019; Chanda and Mukherjee, 2020a, 2020b).

2.6 EVOLUTIONARY CHANGES IN VENOM COMPOSITION AND DELIVERY MECHANISMS IN COBRAS

Analyzing the evolutionary changes in venom composition and delivery mechanisms in Cobras provides valuable insights into their predatory strategies and the selective pressures they have faced over time. Venom is a crucial component of a Cobra's predatory arsenal, allowing them to efficiently immobilize and subdue their prey. In general, Cobras produce a potent venom comprising a myriad of postsynaptic neurotoxins, cytotoxins, cardiotoxins, and enzymes, which induces neurological symptoms and local necrosis (Chu et al., 2010). However, Cobras exhibit remarkable inter-species and intra-species geographical variation in venom composition. Understanding these variations in venom composition (see Chapter 5 for details) and delivery mechanisms among different Cobra species can shed light on the adaptations that have occurred to enhance their hunting efficiency and survival.

Additionally, studying these evolutionary changes can help understand the potential medical applications of Cobra venom and contribute to developing novel therapeutics. For example, researchers have found that the venom of the spitting Cobra species has evolved to immobilize their prey and blind potential predators or threats by accurately spitting venom into their eyes from a distance (Avella et al., 2021; Young et al., 2009). This adaptation has allowed spitting Cobras to effectively defend themselves against larger predators and increase their chances of survival in their natural habitats. Furthermore, studying the unique components of Cobra venom has led to the development of antivenom treatments that can save lives in snakebite cases.

2.7 BEHAVIOURAL EVOLUTION

2.7.1 Behavioural Adaptations in Cobras for Survival

Cobras have developed several behavioural adaptations to enhance their survival. One such adaptation is their ability to hood up, a defensive display used to intimidate predators and potential threats. By spreading their neck flaps and raising their heads, Cobras appear larger and more menacing, deterring many predators from approaching (Jones, 2017). Additionally, Cobras have evolved to be excellent climbers, allowing them to escape danger by ascending trees or other elevated surfaces or swimming quickly (Mukherjee, 2021a). These behavioural adaptations and their venomous spitting ability have made Cobras highly successful predators and survivors in their natural habitats.

Cobras have a remarkable ability to adapt to various environmental conditions, enabling them to thrive in multiple ecosystems. They have also developed sophisticated hunting techniques, such as their ability to strike and inject venom into their prey accurately, even in the dark. Cobras' keen eyesight and ability to detect vibrations further enhance their hunting prowess, allowing them to locate and capture prey precisely (Friedel et al., 2008).

2.7.2 Defensive Mechanisms in Cobras

The evolution of defensive mechanisms can be traced back to millions of years ago when Cobras first began to develop their iconic hoods and venomous fangs. At least three defensive strategies have been identified in Cobras – hooding, hood markings and venom spitting. These adaptations have enabled Cobras to effectively ward off potential predators and efficiently capture and immobilize their prey. The development of such defensive traits has been a crucial factor in the success of Cobras as apex predators, allowing them to maintain their dominance in their respective ecosystems.

The dorsolateral flattening of the body, known as hooding behaviour, serves as an aposematic display, enhancing the snake's appearance to appear larger and more menacing to predators. This behaviour reduces the likelihood of attack. All Asian Cobra species exhibit hooding behaviour achieved through the dorsolateral flattening of the neck region and raising the upper body, creating a more prominent appearance for predators. This characteristic hood is further heightened by elongated thoracic ribs, which are laterally and dorsally rotated to facilitate expansion and dorsoventral compression (Young and Kardong, 2010). The spreading and relaxation of the hood involve a complex interaction of four sets of muscles responsible for lifting the hood, moving the skin, transferring forces between the ribs, and relaxing the hood (Young and Kardong, 2010).

In addition to hooding, Cobras exhibit a behaviour where they elevate the front third of their bodies off the ground while producing their characteristic hood. According to the threat-sensitivity hypothesis proposed by Helfman (1989), prey animals adjust their response based on the perceived threat level of the predator

(Helfman, 1989). Nasoori et al. (2016) discovered a close correlation between the height of vertical posturing, strike frequency, and hissing frequency in *N. oxiana* with the stress induced by the proximity of humans, supporting this hypothesis in Cobras (Nasoori et al., 2016). Moreover, humans have been found to detect snakes in a striking posture significantly faster than those in a resting pose (Masataka et al., 2010). The use of an elevated, ready-to-strike posture by Cobras during defence suggests that predators will readily recognize the heightened threat level.

Specific Cobra species exhibit unique hood markings at the back of the neck, manifesting as a spectacle, monocle, heart, or spot on the hood's rear. These patterns may sometimes extend to the throat area as well. While pigmentation of the pattern sometimes occurs on the scales, it is often primarily found on the interstitial skin between scales. Consequently, the pattern becomes more pronounced when the hood is expanded and the interstitial skin is stretched. These hood patterns are most extensively documented and distinctly observed in Asian species, notably in *Naja naja*, known as the spectacled Cobra, and *Naja kaouthia*, recognized as the monocled Cobra (Figure 2.2). The spectacle marking on the hood in *N. naja* has been hypothesized to mimic eyes akin to the spots on a butterfly (Langerholc, 1991; Stevens, 2005). In a study investigating the responses of Bonnet Macaques to various snake species, Ramakrishnan et al. (2005) discovered that despite Cobras not being predators of Macaques, a model of *N. naja* prompted the quickest reaction time (Ramakrishnan et al., 2005). The recognition of paired eyes is an adaptive trait in primates; therefore, schematic eye spots can be provocative (Emery, 2000). Consequently, an unexpected encounter with the eye-like spots of a Cobra can trigger a startled response, leading to delayed attacks owing to confusion.

While venom primarily serves a predatory function in snakes, spitting venom has evolved solely as a defensive adaptation. When venom enters the eyes of a predator, it induces immediate pain and can result in temporary or permanent blindness if left untreated (Chu et al., 2010). Thus, this defensive mechanism serves as an instant defence against predators. The venom-spitting projectile defence mechanism is

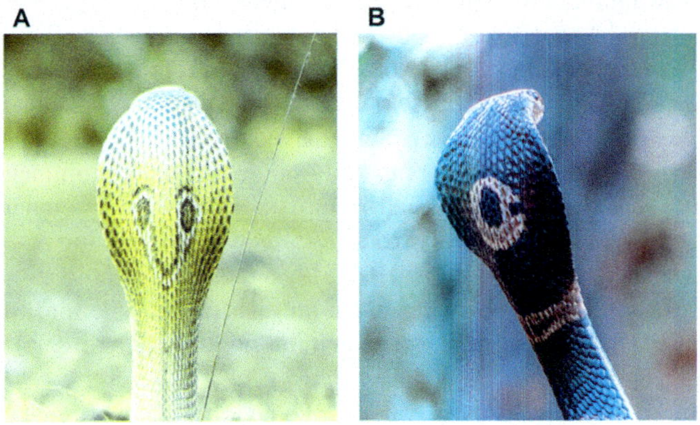

FIGURE 2.2 Distinct hood markings of **A**. *Naja naja* and **B**. *Naja kaouthia*.

(Photo courtesies: Mr. Subhadeep Dey.)

prevalent in seven Asian Cobra species, including *N. atra, N. sumatrana, N. siamensis, N. sputatrix, N. philippinensis, N. mandalayensis* and *N. samarensis*.

2.8 EVOLUTIONARY SIGNIFICANCE OF ADAPTATIONS AND THEIR CONTRIBUTION TO COBRAS' SURVIVAL AND REPRODUCTIVE SUCCESS

The evolution of the iconic hood and venomous fangs in Cobras has resulted from natural selection favouring individuals with these traits. Cobras with these defensive mechanisms were more likely to survive encounters with predators and were better equipped to catch and subdue their prey. As a result, they were more likely to pass on their genetic traits to the next generation, increasing the prevalence of these adaptations over time. This evolutionary advantage has allowed Cobras to maintain their position as top predators in their ecosystems, ensuring their continued survival and success.

Regarding reproduction, Cobras engage in courtship rituals that involve intricate dances and displays of strength. Males compete for the attention of females, often engaging in fierce battles to establish dominance. Once a female selects a mate, she lays eggs and diligently guards them until they hatch (Mukherjee, 2021a). Cobras' survival and reproductive strategies have ensured their continued existence and dominance in their ecosystems.

2.9 THE GENETIC VARIABILITY AND DIVERSITY WITHIN THE COBRA SPECIES

Genetic evolution is crucial in shaping Cobra species' genetic variability and diversity (Ratnarathorn et al., 2023). Through genetic mutations and recombination, Cobras have developed various adaptations that aid their survival and reproduction. These adaptations can range from changes in their venom composition to differences in colouration and body size (Lin et al., 2008; Mukherjee, 2021a). For instance, Lin and their colleagues studied the quantity of black pigment on the ventral scales of Chinese Cobras (*Naja atra*) in Taiwan. They discovered significant variations in ventral colouration among different populations. Research has shown that the populations in eastern Taiwan possess black ventral scales, while populations in the western region are primarily white. The nucleotide diversity in the mitochondrial regulatory region sequences is shallow. Snakes with black and white ventral colouration have significant similarities in their genetic makeup and do not exhibit any genetic variation. According to the findings of their research, Lin et al. (2008) concluded that ventral colouring may be a consequence of local adaptation. Therefore, the genetic variability within the Cobra species allows them to adapt to different environments and prey, increasing their chances of survival in a changing world (Chang et al., 2004).

In 2020, Kundu and colleagues employed the DNA barcoding approach, specifically utilizing information from the cytochrome c oxidase subunit I (COI) gene, to investigate the population genetic diversity of lethal Cobra species (*N. naja* and *N. kaouthia*) in the Mizoram State in NE India and Rangpur in the northern Bangladesh region. As shown in Figure 2.3, the molecular-based study on *N. kaouthia* reveals the

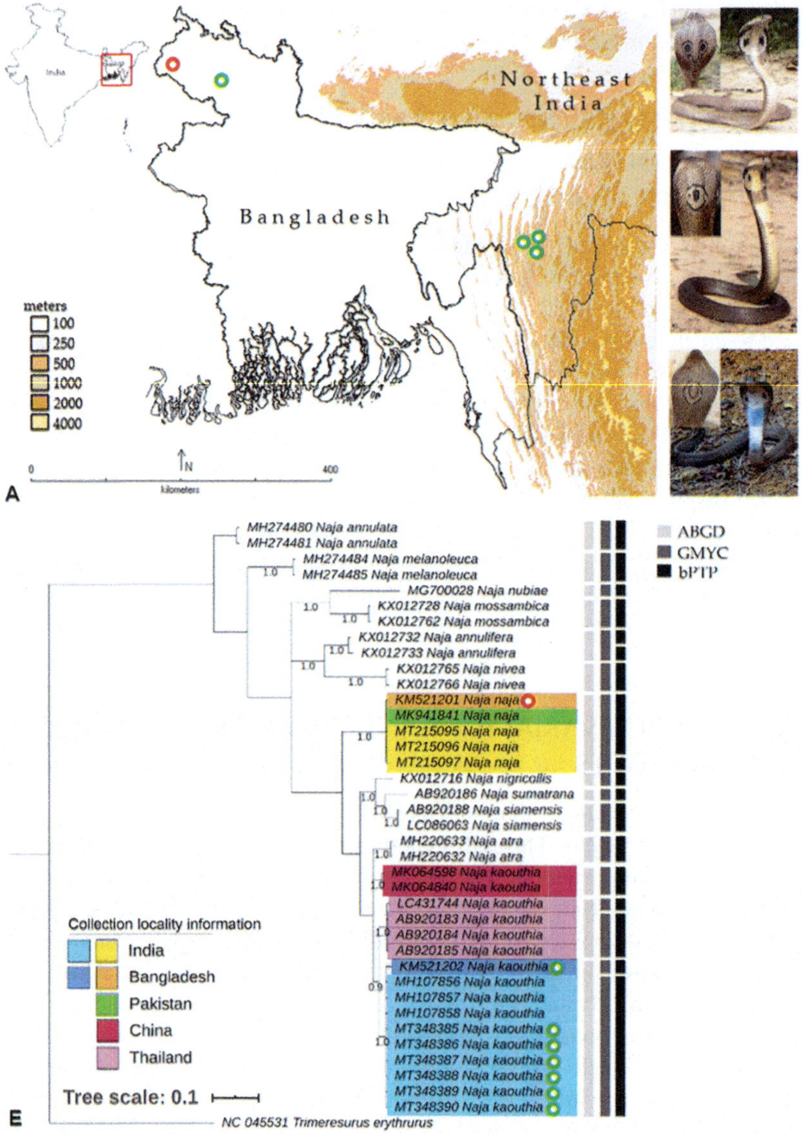

FIGURE 2.3 (A) Collection locality map of *N. naja* and *N. kaouthia* from the Indo-Bangladesh region. (B-D) Live photographs of *N. naja* and *N. kaouthia* collected from the Indo-Bangladesh region, (E) Bayesian analysis based on partial mtCOI (mitochondrial cytochrome c oxidase subunit I) gene inferred phylogenetic relationship of *Naja* species. The clades of *N. naja* and *N. kaouthia* are marked by different colours per their collection locality information. The generated sequences are marked by different colour dots as used in the collection locality map. Grey shaded colour bars beside the tree species delimitation methods indicate delineated MOTUs (Molecular Taxonomic Units) by different species delimitation methods (ABGD, Automatic Barcode Gap Discovery; GMYC, General Mixed Yule-Coalescent; bPTP, Poisson-Tree-Processes).

(Reproduced from Kundu et al., 2020 (original Figure 1a–e) under open access licence (http://creativecommons.org/licenses/by-nc/4.0/).)

existence of hidden diversity in Indo-Bangladesh, China, and Thailand by the combination of different species delimitation methods, reciprocal monophyly requirements, and genetic distance (Kundu et al., 2020; Ratnarathorn et al., 2019). Therefore, it may be concluded that genetic mutations are crucial in driving evolutionary change in Cobras. These mutations, which are random changes in the DNA sequence, can lead to new traits or variations in existing traits.

Another study by Lin and co-workers aimed to understand the population structure and genetic variability among *N. atra* snakes from Mainland China and Vietnam. They genotyped 12 microsatellite loci and sequenced 1117 bp of the mitochondrial cytochrome b gene from 269 specimens. Results indicated significant genetic diversity in both mitochondrial DNA and microsatellite loci. The mitochondrial DNA analysis identified two primary clades: one encompassing Vietnam, southern China, and southwestern China, and another consisting of eastern and southeastern China. A minor clade, represented by two individuals from the westernmost site, was also identified. Microsatellite data further divided the eastern and southeastern China clades into two distinct genetic clusters, corresponding to these regions. The authors hypothesized that geographical features of the Luoxiao and Nanling mountains may have played a crucial role as barriers influencing lineage diversification (Lin et al., 2012). Notably, these minor variations in the population can provide an advantage in certain environments or when hunting specific prey, thereby increasing the individual's chances of survival and reproductive success.

2.10 CONCLUSION: COBRAS ARE THREATENED SPECIES OF SNAKES, AND IT IS CRUCIAL TO CONSERVE THEM

Given the increasing human populations, it is crucial to accurately map the current and future territories of Asian Cobras for biological conservation and health care administration. Analyses indicate that Cobras living in arid climates and on islands have specialized and limited ranges of suitable climate conditions. In contrast, Cobras found on the mainland with more significant geographic distributions have broader ranges of suitable climate conditions (Chowdhury et al., 2021). Additionally, as shown in Figure 2.4, their research revealed a substantial increase in the fragmentation of anticipated Cobra distributions. Over the next 50 years, Asian Cobras are projected to experience a loss of around 60% of their existing climatic range, which is suited for their survival. It has been suggested that at some point in time, *Naja mandalayensis*, *N. sputatrix*, *N. samarensis*, and *N. philippinensis* will probably have no remaining climate space that is suitable and accessible. In addition, a continued rise in populations of humans in this area could significantly amplify the consequences of human-caused influences (Chowdhury et al., 2021).

From a conservation standpoint, the primary peril to Cobras is altering habitats (IUCN, 2019). Overall, the evolutionary history of Cobras serves as a valuable case study for understanding the intricate interplay between adaptation, biodiversity, and species conservation. Moreover, the research demonstrates that Asian Cobras are endangered due to the impacts of climate change and the increasing human population (Chowdhury et al., 2021). Climate change is expected to decrease the size of the habitats and the range of the Asian mainland Cobras (Figure 2.5). This drastic

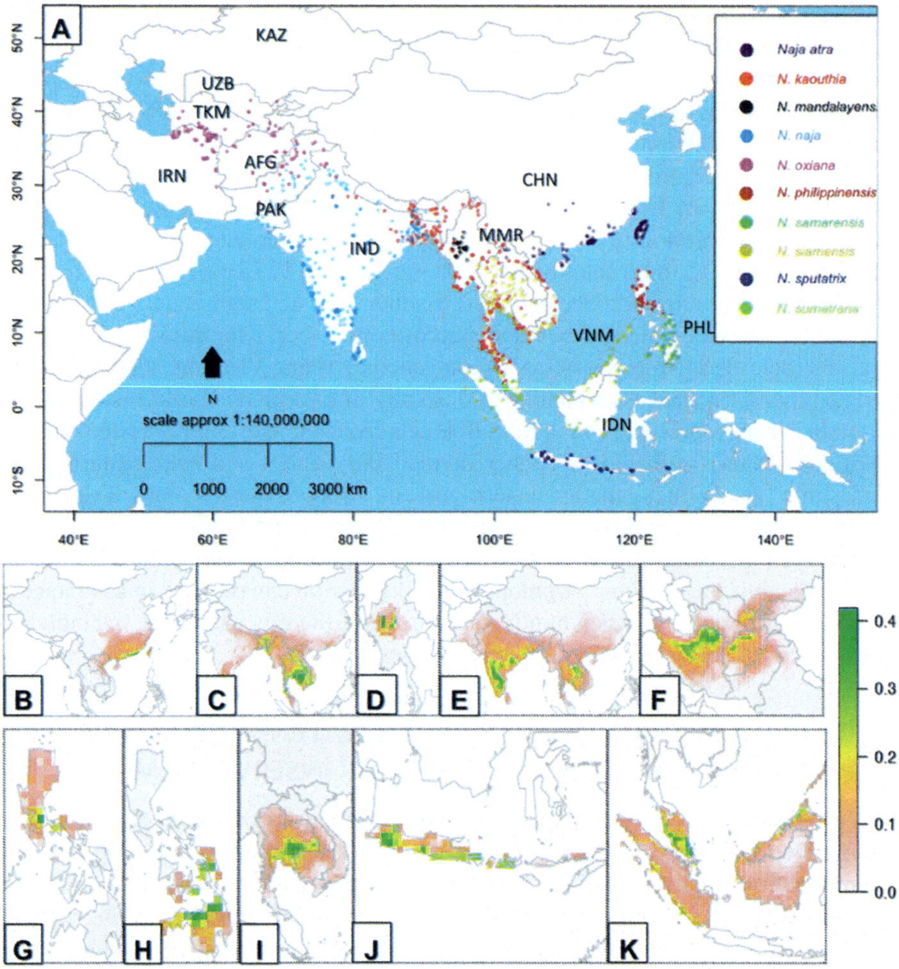

FIGURE 2.4 (A) The occurrence points of ten Asian Cobras used in training the climate prediction model. The current potential climatic niche of ten Asiatic Cobras that were produced by consensus of CCSM (Community Climate System Model), CNRM (Centre National de Recherches Météorologiques), MIROC (Model for Interdisciplinary Research on Climate), and MIR (Meteorological Research Institute) global climate models (B) *Naja atra*, (C) *N. kaouthia*, (D). *N. mandalayensis*, (E). *N. naja*, (F) *N. oxiana*, (G). *N. philippinensis*, (H) *N. samarensis*, (I) *N. siamensis*, (J) *N. sputatrix*, and (K) *N. sumatrana*.

(Reproduced from Chowdhury et al., 2021 (original Figure 1a-k) under open access licence (http://creativecommons.org/licenses/by-nc/4.0/).)

climate change would result in a shortage of suitable habitats for these snakes within a few decades.

Furthermore, the swift alteration of land use will also affect habitats that may be reached or entered. Southeast Asia is a suitable region for five species of Cobras. However, it is expected that due to the rapid change of land and excessive

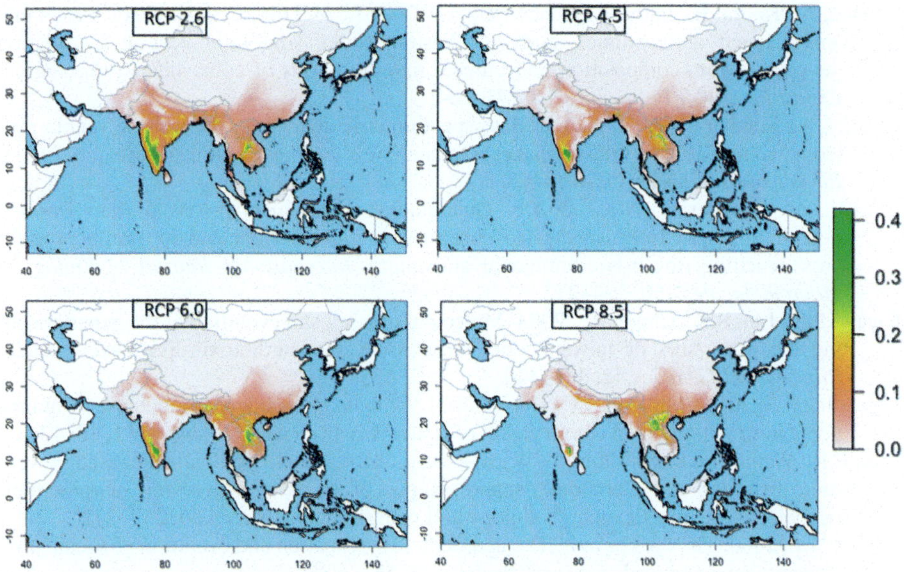

FIGURE 2.5 The consensus map of the future (2070) suitable climatic niche of *Naja naja* in four different Representative Concentration Pathway (RCP) trajectories. The largest suitable climatic space is RCP 2.6, and the least is RCP 8.5.

(Reproduced from Chowdhury et al., 2021 (original Figure 2) under open access licence (http://creativecommons.org/licenses/by-nc/4.0/).)

exploitation, there will be a significant increase in the deterioration of their native environments (Chowdhury et al., 2021).

Therefore, the conservation of Cobras is crucial to ensure their long-term survival and evolutionary potential. Efforts must be focused on implementing conservation strategies such as habitat protection, captive breeding programs, and reducing human-induced pressures. Furthermore, enhancing surveillance and the presence of local law enforcement agencies could limit commerce, enhance public consciousness, and mitigate excessive abuse. Additionally, addressing the potential genetic bottlenecks and loss of gene diversity in Cobra populations is essential to maintain their ability to adapt to changing environments and overcome future challenges. By prioritizing the conservation of these iconic reptiles, we can ensure a brighter future for Cobras and the ecosystems they inhabit.

REFERENCES

Avella, I., Barajas-Ledesma, E., Casewell, N.R., Harrison, R.A., Rowley, P.D., Crittenden, E., Wüster, W., Castiglia, R., Holland, C., van der Meijden, A., 2021. Unexpected lack of specialisation in the flow properties of spitting cobra venom. *Journal of Experimental Biology* 224, jeb229229.

Ceriaco, L.M., Marques, M.P., Schmitz, A., Bauer, A.M., 2017. The "cobra-preta" of São Tomé island, gulf of Guinea, is a new species of *Naja laurenti*, 1768 (Squamata: Elapidae). *Zootaxa* 4324, 121–141.

Chanda, A., Kalita, B., Patra, A., Senevirathne, W.D.S.T., Mukherjee, A.K., 2019. Proteomic analysis and antivenomics study of Western India *Naja naja* venom: Correlation between venom composition and clinical manifestations of cobra bite in this region. *Expert Review of Proteomics* 16, 171–184.

Chanda, A., Mukherjee, A.K., 2020a. Mass spectrometric analysis to unravel the venom proteome composition of Indian snakes: Opening new avenues in clinical research. *Expert Review of Proteomics* 17, 411–423.

Chanda, A., Mukherjee, A.K., 2020b. Quantitative proteomics to reveal the composition of Southern India spectacled cobra (*Naja naja*) venom and its immunological cross-reactivity towards commercial antivenom. *International Journal of Biological Macromolecules* 160, 224–232.

Chang, L.S., Lin, S.K., Chung, C., 2004. Molecular cloning and evolution of the genes encoding the precursors of taiwan cobra cardiotoxin and cardiotoxin-like basic protein. *Biochemical Genetics* 42, 429–440.

Chowdhury, M.A.W., Muller, J., Varela, S., 2021. Climate change and the increase of human population will threaten conservation of Asian cobras. *Scientific Reports* 11, 18113.

Chu, E.R., Weinstein, S.A., White, J., Warrell, D.A., 2010. Venom ophthalmia caused by venoms of spitting elapid and other snakes: Report of ten cases with review of epidemiology, clinical features, pathophysiology and management. *Toxicon* 56, 259–272.

Emery, N.J., 2000. The eyes have it: The neuroethology, function and evolution of social gaze. *Neuroscience and Biobehavioral Reviews* 24, 581–604.

Friedel, P., Young, B.A., van Hemmen, J.L., 2008. Auditory localization of ground-borne vibrations in snakes. *Physical Review Letters* 100, 048701.

Helfman, G., 1989. Threat-sensitive predator avoidance in damselfish-trumpetfish interactions. *Behavioral Ecology and Sociobiology* 24, 47–58.

IUCN, 2019. The IUCN red list of threatened species, Version 2019-3.

Jackson, K., 2003. The evolution of venom-delivery systems in snakes. *Zoological Journal of the Linnean Society* 137, 337–354.

Jones, B., 2017. *The evolution of defensive strategies in Cobras*. Bangor University (United Kingdom).

Kardong, K.V., 1982. The evolution of the venom apparatus in snakes from colubrids to viperids and elapids. *Memórias do Instituto Butantan* 46, 105–118.

Kardong, K.V., Weinstein, S., Smith, T., 2009. Reptile venom glands: Form, function, and future. *Handbook of venoms and toxins of reptiles*, CRC Press, pp. 65–91.

Kundu, S., Lalremsanga, H.T., Rahman, M.M., Ahsan, M.F., Biakzuala, L., Kumar, V., Chandra, K., Siddiki, A.Z., 2020. DNA barcoding elucidates the population genetic diversity of venomous cobra species (Reptilia: Elapidae) in Indo-Bangladesh region. *Mitochondrial DNA Part B* 5, 2525–2530.

Langerholc, J., 1991. Facial mimicry in the animal kingdom. *The Italian Journal of Zoology* 58, 185–204.

Lin, H.-C., Li, S.-H., Fong, J., Lin, S.-M., 2008. Ventral coloration differentiation and mitochondrial sequences of the Chinese Cobra (*Naja atra*) in Taiwan. *Conservation Genetics* 9, 1089–1097.

Lin, L.-H., Qu, Y.-F., Li, H., Zhou, K.-Y., Ji, X., 2012. Genetic structure and demographic history should inform conservation: Chinese cobras currently treated as homogenous show population divergence. *PLoS One* 7, e36334.

Logan, J.M., Beck, A., Dunstan, N., Allen, L., Woods, A.E., 2021. Development of the venom delivery system in Elapidae snake species: *Naja siamensis* and *Oxyuranus microlepidotus*. *Toxicon* 199, 12–19.

Marais, J., 2022. *A complete guide to the snakes of southern Africa*. Penguin Random House South Africa.

Masataka, N., Hayakawa, S., Kawai, N., 2010. Human young children as well as adults demonstrate 'superior'rapid snake detection when typical striking posture is displayed by the snake. *PLoS One* 5, e15122.

Mukherjee, A.K., 2020. Species-specific and geographical variation in venom composition of two major cobras in Indian subcontinent: Impact on polyvalent antivenom therapy. *Toxicon* 188, 150–158.

Mukherjee, A.K., 2021a. *Indian spectacled cobra (Naja naja), the 'Big Four' snakes of India: Venom composition, pharmacological properties and treatment of envenomation*. Springer, pp. 69–94.

Mukherjee, A.K., 2021b. *Evolution of snakes and systematics of the "Big Four" venomous snakes of India, The 'Big Four' snakes of India: Venom composition, pharmacological properties and treatment of envenomation*. Springer, pp. 21–34.

Nasoori, A., Shahbazzadeh, D., Tsubota, T., Young, B.A., 2016. The defensive behaviour of *Naja oxiana*, with comments on the visual displays of cobras. *Herpetological Bulletin* 138, 13–17.

Panagides, N., Jackson, T.N., Ikonomopoulou, M.P., Arbuckle, K., Pretzler, R., Yang, D.C., Ali, S.A., Koludarov, I., Dobson, J., Sanker, B., 2017. How the cobra got its flesh-eating venom: Cytotoxicity as a defensive innovation and its co-evolution with hooding, aposematic marking, and spitting. *Toxins* 9, 103.

Ramakrishnan, U., Coss, R.G., Schank, J., Dharawat, A., Kim, S., 2005. Snake species discrimination by wild bonnet macaques (*Macaca radiata*). *Ethology* 111, 337–356.

Ratnarathorn, N., Harnyuttanakorn, P., Chanhome, L., Evans, S.E., Day, J.J., 2019. Geographical differentiation and cryptic diversity in the monocled cobra, *Naja kaouthia* (Elapidae), from Thailand. *Zoologica Scripta* 48, 711–726.

Ratnarathorn, N., Nadolski, B., Sumontha, M., Hauser, S., Suntrarachun, S., Khunsap, S., Laoungbua, P., Radcliffe, C.A., Vasaruchapong, T., Tawan, T., 2023. An expanded description, natural history, and genetic variation of the recently described cobra species *Naja fuxi* Shi et al., 2022. *Vertebrate Zoology* 73, 257–276.

Senji Laxme, R.R., Attarde, S., Khochare, S., Suranse, V., Martin, G., Casewell, N.R., Whitaker, R., Sunagar, K., 2021. Biogeographical venom variation in the Indian spectacled cobra (*Naja naja*) underscores the pressing need for pan-India efficacious snakebite therapy. *PLoS Neglected Tropical Diseases* 15, e0009150.

Stevens, M., 2005. The role of eyespots as anti-predator mechanisms, principally demonstrated in the Lepidoptera. *Biological Reviews* 80, 573–588.

Wallach, V., Wuester, W., Broadley, D.G., 2009. In praise of subgenera: Taxonomic status of cobras of the genus *Naja laurenti* (Serpentes: Elapidae). *Zootaxa* 2236, 26–36.

Wüster, W., 1996. Taxonomic changes and toxinology: Systematic revisions of the Asiatic cobras (*Naja naja* species complex). *Toxicon* 34, 399–406.

Wüster, W., Crookes, S., Ineich, I., Mané, Y., Pook, C.E., Trape, J.-F., Broadley, D.G., 2007. The phylogeny of cobras inferred from mitochondrial DNA sequences: Evolution of venom spitting and the phylogeography of the African spitting cobras (Serpentes: Elapidae: *Naja nigricollis* complex). *Molecular Phylogenetics and Evolution* 45, 437–453.

Young, B.A., Boetig, M., Westhoff, G., 2009. Functional bases of the spatial dispersal of venom during cobra "spitting". *Physiological and Biochemical Zoology* 82, 80–89.

Young, B.A., Kardong, K.V., 2010. The functional morphology of hooding in cobras. *Journal of Experimental Biology* 213, 1521–1528.

Zhu, X., Zhu, G., Zhang, S., Du, Y., Qu, Y., Lin, L., Ji, X., 2022. Lineage Diversification and Niche Evolution in the Chinese Cobra *Naja atra* (Elapidae). *Asian Herpetological Research* 13, 242–250B.

3 Venom Delivery System in Asian Cobras

3.1 VENOM GLAND OF COBRAS

Three families of venomous snakes arose along the course of evolution – Colubridae, Viperidae, and Elapidae. Each of these groups developed their fangs and venom delivery systems independently during evolution, resulting in subtle changes in the anatomy of their fangs to grasp prey tightly in the mouth. The first two groups (Viperidae and Elapidae) developed a sophisticated venom system to kill prey.

Snake venom glands are modified parotid salivary glands that secrete toxins via modified saliva or spit. The venom glands are on the side of the upper jaw, behind the eyes. Venomous snakes, including cobras, gradually produce venom during hibernation (winter sleep), and because the venom serves no purpose at this time, it accumulates in venom glands.

On either side of their heads, Cobras have an oval-shaped venom gland. The venom gland comprises an extended mucous accessory gland in the front area surrounding a secretory duct and a primary gland at the posterior end (Bdolah, 1979; Kardong et al., 2009; Rosenberg, 1967). The gland's hind end is turned ventral in some species, giving rise to a noticeable bulge, whereas in others, there is just a slight protrusion. At its anterior end, the venom gland tapers at the level of the postorbital bone, and a relatively narrow segment runs ventral to the orbit (Figure 3.1). The supplementary mucous gland encircles this little segment, the main gland duct (Rosenberg, 1967).

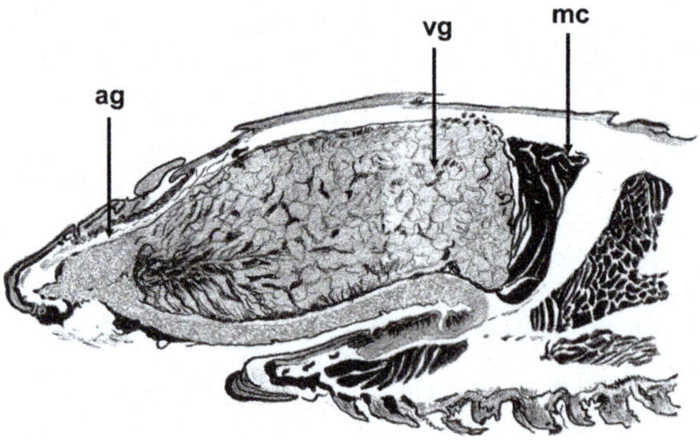

FIGURE 3.1 Venom gland of the cape Cobra. ag - accessory gland; mc - compressor muscle; vg - main venom gland.

(Redrawn from Kochva et al., 1982.)

DOI: 10.1201/9781003598121-3

Histological investigations reveal many simple or branching ducts in the primary gland. There isn't a central lumen since these ducts converge anteriorly toward the secretory duct, temporarily releasing venom through the accessory glands (Jackson, 2003; Kochva et al., 1982; Rosenberg, 1967). A dense capsule of connective tissue surrounds the venom and auxiliary mucous glands. The venom apparatus's muscles, ligaments, and connective tissues are linked to this capsule's exterior (Rosenberg, 1967).

The primary cell type in charge of creating, secreting, and storing venom is secretory epithelial cells, which are often aligned with the tubules of the venom gland. The cells in question consist of a basal nucleus, a rough endoplasmic reticulum, and secretory granules (Mackessy, 2022).

Compressor glandulae are unique muscle fibers that connect the venom gland's lateral and medial posterior areas. These muscle fibers aid in the venom's expulsion from the primary gland, which allows it to enter the fang through the accessory gland (Mackessy, 2022; Rosenberg, 1967).

3.2 VENOM DELIVERY SYSTEM IN COBRAS

Cobra is a member of the front-fanged venom delivery system's proteroglyph system. This delivery system has a unique tooth called a fang rostrally connected to the maxilla's mouth. Via a suture joining the entrance and exit pores, the fangs are tubular structures that allow the venom to be administered to a victim or prey (Deufel and Cundall, 2006).

The functioning fangs are located in two neighboring fang sockets on the rostral end of the maxilla. When viewed ventrally, the maxilla assumes a C-shaped appearance, with both ends appearing to bend medially. From a lateral perspective, on the other hand, the maxilla seems to be taller rostrally. The maxillary-prefrontal junction is just below or somewhat rostral to the fang (Figure 3.2). According to Deufel and Cundall (2003), the maxilla and prefrontal are widely connected, enabling a rocking-sliding motion during fang erection. The little ligament fibers regulate teeth rotation and traverse the large joint between the maxilla and the prefrontal (Deufel and Cundall, 2003, 2006).

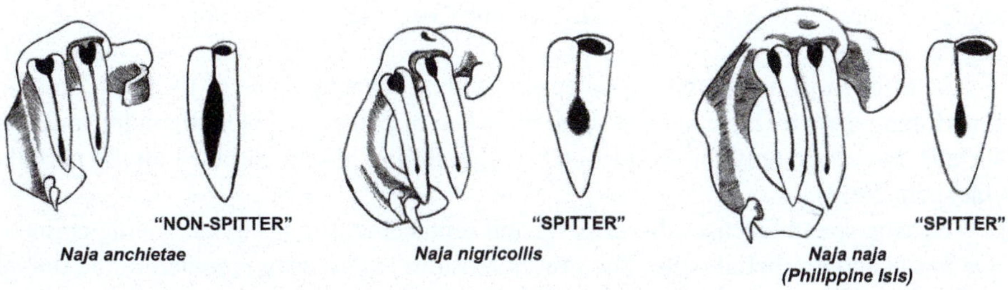

FIGURE 3.2 Ventral and anterior views of the maxillae and distal portion of a fang, respectively.

(Redrawn from Bogert, 1943.)

Spitting Cobras differ from non-spitting cobras due to several morphological modifications in their venom delivery system, which are crucial for self-defense. After a detailed investigation in 1943, Bogert found variations between non-spitting and spitting Cobras' fangs. Non-spitters' venom ducts have a seam that runs from the entrance lumen to the discharge pore. The discharge lumen emerges at the outside edge of a non-spitter suture as a shallow groove in front of the pointed portion. This suture swings slightly outward. In contrast, the suture in the case of spitters is located about on the midline of the anterior area of the curve (Bogert, 1943).

The discharge pore in non-spitting Cobras has an extended aperture. Nevertheless, it resembles a teardrop in spitters, with the bottom edge seeming rounded and the upper edge terminating where the venom canal's walls converge at the suture. Because of this modification, during an assault, the venom can be released forward rather than downward. Therefore, the venom makes a dramatic 90° rotation before exiting the tooth (Bogert, 1943; Triep et al., 2013; Wüster and Thorpe, 1992).

The presence of ridges in the venom channel is another distinctive characteristic of the teeth of the spitting Cobra. These typically paired ridges stretch from the distal end of the venom channel to its last third. They terminate just before an abrupt 90° turn that reroutes flow in the direction of the ejection orifice (Triep et al., 2013; Young et al., 2009). Because of all these morphological changes, the spitting Cobra can shoot its venom to a great distance.

3.3 FANGS OF COBRAS

The maxilla of a Cobra often has tiny, tubular fangs. Usually, there are two asymmetric canine teeth on each maxilla. While the other tooth is non-functional and only partially fused, one is fully connected and functional. Consequently, accessible alternate canine teeth occupy each maxilla's internal and external sockets (Berkovitz and Shellis, 2016).

Each replacement dentition has at least two or three behind it, depending on its developmental stage. The functional fang separates from the growing fang as it approaches completion, and the developing fang's base enters the empty socket while staying horizontal (Jackson, 2007).

The replacement fang rotates from its initial horizontal position to its functional role, where the maxilla's horizontal and central axes are inclined at a 30- to 40-degree angle (Figures 3.3–3.4). Eventually, it ankyloses (Berkovitz and Shellis, 2016; Bogert, 1943).

The three most noticeable replacement teeth at the apical end of the tooth and the functioning teeth all have a full discharge orifice of comparable length on the anterior surface. However, only the ankylosed functional fang has an entrance on the maxilla (Jackson, 2002).

The fang sheath encases the original and replacement fangs by enclosing connective tissue and epithelial cells. The oral mucosa in the vicinity is joined to the dorsal edge of the fang sheath by the lateral and cranial surfaces of the maxilla. The venom duct travels through the fang sheath's caudolateral surface before continuing its route via the craniad (Young et al., 2004).

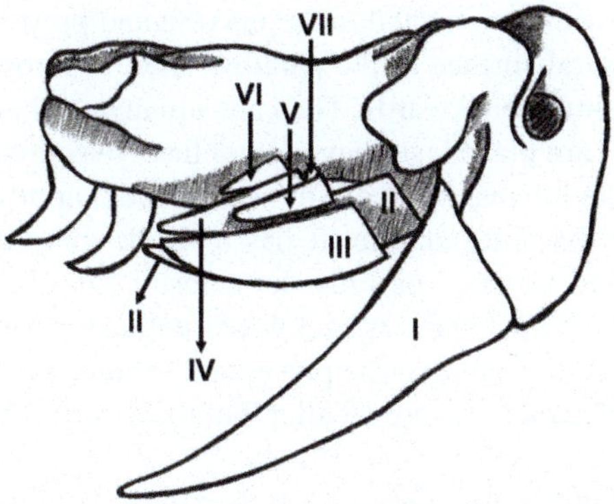

FIGURE 3.3 *Naja melanoleuca's* left maxilla and fangs in the replacement series. The functioning fang is represented by the Roman number I. The additional Roman numbers represent the teeth as they grow to full size, become stiffened (ankylosed), and become functional.

(Redrawn from Bogert, 1943.)

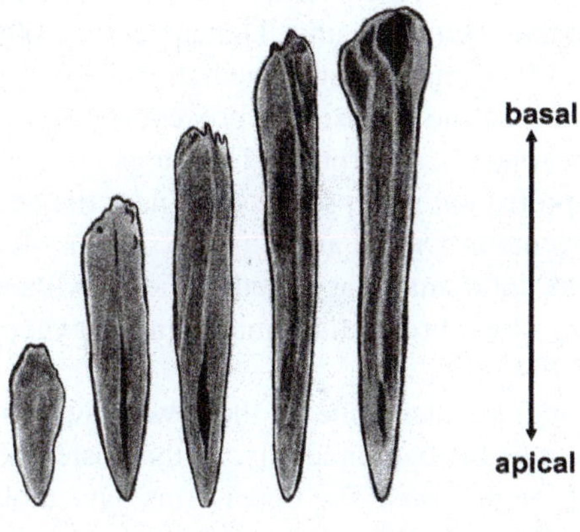

FIGURE 3.4 Anterior image of a working and replacement fang series from a woodland Cobra (*Naja melanoleuca*).

(Redrawn from Jackson, 2002.)

3.4 DENTITION

Cobras have multiple tooth-bearing bones in their upper and lower jaws. The upper jaw, known as the palato-maxillary arch, is made up of four bone types: palatine, pterygoid, ectopterygoid, and maxilla, whereas the lower jaw is made up of dentary bones.

The palatine is an extended bone that runs parallel to and medial to the maxilla. The palatine bone's back end is next to the pterygoid bone's front tip. The palatine bone continues craniolaterally from this location and approaches the cranial tip of the maxilla bone. The palatine bone's cranial end (front end) supports the extensive medial process, which is attached to the frontal bone by connective tissue. It also supports the lateral process connected to the maxilla and prefrontal bones by connective tissue. On the ventral surface of the palatine, there is a row of small conical and curved sharp teeth pointing backwards. They are situated in shallow slots where small portions of their bases are placed, so they are not firmly secured (Kamal et al., 1970).

The pterygoid is the longest bone on the ventral side, approximately horizontally aligned with the maxilla and palatine. It has a slight curvature in its anterior and posterior sections when viewed ventrally. Its anterior fourth is cylindrical and thin, while the remainder gradually spreads backwards with a tapering posterior extremity. Small conical teeth can be found on the pterygoid's anterior two-thirds ventromedial surface. The front and back teeth are smaller, but the other teeth are all the same size (Kamal et al., 1970).

The ectopterygoid bones are craniolateral in extension with a minor dorsal deflection. It lacks teeth, unlike the other jaw bones. It has a poorly defined extended articulation on the dorsal side of the pterygoid at its proximal end and a broad, horizontal plate of bone at its distal end (Young et al., 2004).

The maxilla bone's caudal end sits beneath the ectopterygoid bone's distal end. Dense connective tissue fills the medial region of the contact between the maxilla and ectopterygoid bones. Conversely, the lateral section has a more complicated articulation with little connective tissue (Young, Dunlap et al. 2004). The maxilla has a medial palatine process that arises posteromedial to the two fang sockets and coils ventrally to connect the palatine's maxillary process on the dorsal side. The maxilla protrudes anteriorly beyond the prefrontal. The fangs are somewhat anterior to the axis of the maxillary-prefrontal joint, so they are near the center of maxillary rotation, limiting their movement (Deufel and Cundall, 2003). Many Cobra species typically have one or two solid maxillary teeth. The maxillary teeth are short and separated from the fangs by a broad diastema (a gap between teeth) (Berkovitz and Shellis, 2016; Bogert, 1943).

The dentary is a long, slender bone of the lower jaw. The anterior end has an upward curve and is somewhat bent medially. At the posterior end, the dentary bone divides into two equal-length arms. The upper arms have a blunt end, but the lower arms taper gradually at the back side and hug the anterior end of the composite bone. A broad, conical, firmly pointed, and posteriorly directed row of teeth in shallow grooves runs along the whole upper border of dentary bones. These teeth gradually shrink in the posterior direction (Kamal et al., 1970).

3.5 BITING MECHANISM OF COBRAS

The skull and jaw bones are incredibly malleable and instead loosely articulated. This characteristic offers excellent flexibility when striking and/or swallowing. The fangs of Cobras are permanently erected.

The three types of specialized muscles are linked to the venom glands and play an essential role in biting (Figure 3.5). These are as follows:

1. Digastric Muscle: This muscle is linked at one end to the squamosal bone of the skull and at the other end to the articular bone of the lower jaw. The lower jaw is depressed by the contraction of digastric muscles, which facilitates mouth opening.

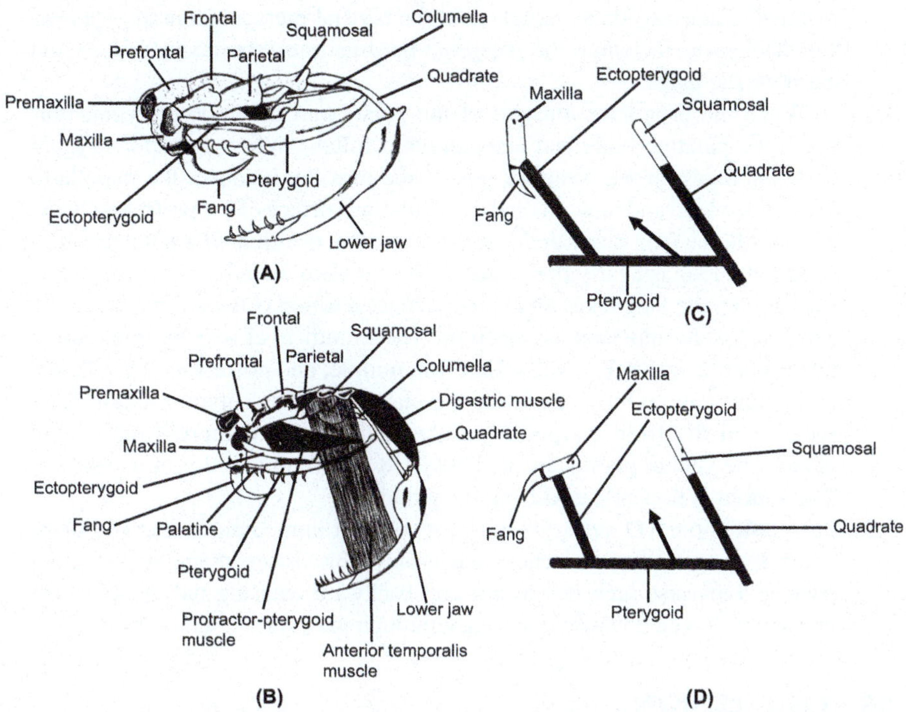

FIGURE 3.5 A partially open (A) and fully open (B) mouth. Schemes depicting the relative arrangement of the major bones involved in fang erection (C and D).

(Redrawn from Jackson, 2002.)

2. Sphenopterygoid: Also known as the protactor-pterygoid muscle. It is linked anteriorly to the spherical region and posteriorly to the dorsal side of the pterygoid. It assists in propelling the pterygoid forward.
3. The temporalis anterior and posterior: It is linked to the lateral walls of the skull and the lower jaw. These muscles are responsible for closing the jaw.

Several coordinated actions are involved in a snake's biting procedure. The striking of the prey, the opening of the mouth and elevation of the fangs, the closure of the jaws followed by venom injection into the prey, and lastly, retraction of the fangs are the four separate phases of these dramatic motions (Figure 3.5).

The following are the descriptions of the biting mechanism in Cobra:

i. The Strike: In the first phase, the snake vigorously and rapidly throws itself forward toward its prey, covering a distance about one-third of its length. Cobras usually raise their heads from the ground to compensate for the limited mobility of fangs.
ii. Opening of the mouth and elevation of the fangs: The Cobra's mouth remains closed before striking, but as the head approaches its prey, the digastric, cervico-mandibular, and vertebra-mandibular muscles contract swiftly, depressing the

mandibles. Contraction of the sphenoid and parietal-pterygoid muscles happens simultaneously, swinging the pterygoid-palatine-transverse arch forward and elevating the fangs.

When the protractor muscles of the palate pull the endo-pterygoid forward, the palatine bone and ectopterygoid follow. When the ectopterygoid bone impinges on the posterior arm of the maxilla, it causes the maxilla to migrate forward and upward on the articulating surface of the prefrontal bone. As a result, the fang raises and assumes a striking position. When the mouth is closed, the fang is nearly horizontal, whereas when the mouth is opened during the bite, the fang takes an almost vertical position (Figure 3.5).

iii. Mouth closure and venom injection: The mouth is closed by contracting three sets of temporal muscles (anterior, middle, and posterior). These muscles collaborate to raise or elevate the mandibles. The superior and inferior sections of the front temporal muscles additionally compress the venom gland. The venom gland capsule is twisted or torn due to this compression. The venom is thus released from the gland.

iv. Fang retraction: The retractor muscles contract immediately after the snake inserts its fangs and injects venom into its prey. This process drags the pterygoid-palatine-transverse arch downward and backward, causing the initially high fangs to be forced downward and back through the tissues (Fairley, 1934).

3.6 CONCLUSION

In conclusion, with their distinctive hood used as a threat display, Cobras stand out as one of Asia and Africa's most notorious venomous snakes. They belong to the Elapidae family of snakes, and have developed sophisticated venom delivery systems independently, alongside viperids, to subdue prey effectively. The evolutionary adaptations in Cobras have led to subtle anatomical variations in their fangs, enhancing their predatory capabilities. The venom delivery system comprises a postorbital venom gland on both sides of the upper jaw, linked to specialized fangs or teeth injecting venom to prey or victim. The venom gland and fangs are hypothesized to have evolved from the same ancestor and are developmentally related. Like other snakes of medical importance to humans, Cobras have front-fanged bodies, which evolved separately. A snake's biting method consists of several coordinated moves. The four distinct phases of these spectacular gestures are hitting the victim, opening the mouth and elevating the fangs, closing the jaws followed by venom injection into the prey, and finally retracting the fangs. The venom apparatus may be primarily for killing or paralyzing prey or may be a purely defensive adaptation. Some venoms also function as digestive fluids.

REFERENCES

Bdolah, A., 1979. *The venom glands of snakes and venom secretion, snake venoms*. Springer, pp. 41–57.

Berkovitz, B., Shellis, P., 2016. *The teeth of non-mammalian vertebrates*. Academic Press, pp. 235–254.

Bogert, C.M., 1943. Dentitional phenomena in cobras and other elapids, with notes on adaptive modifications of fangs. *Bulletin of the AMNH* 81, 3.

Deufel, A., Cundall, D., 2003. Prey transport in "palatine-erecting" elapid snakes. *Journal of Morphology* 258, 358–375.

Deufel, A., Cundall, D., 2006. Functional plasticity of the venom delivery system in snakes with a focus on the poststrike prey release behavior. *Zoologischer Anzeiger-A Journal of Comparative Zoology* 245, 249–267.

Fairley, N.H., 1934. Snake bite: Its mechanism and modern treatment: (Section of tropical diseases and parasitology). *Proceedings of the Royal Society of Medicine* 27, 1083–1094.

Jackson, K., 2002. How tubular venom-conducting fangs are formed. *Journal of Morphology* 252, 291–297.

Jackson, K., 2003. The evolution of venom-delivery systems in snakes. *Zoological Journal of the Linnean Society* 137, 337–354.

Jackson, K., 2007. The evolution of venom-conducting fangs: Insights from developmental biology. *Toxicon* 49, 975–981.

Kamal, A., Hammonda, H., Mokhtar, F., 1970. The development of the osteocranium of the Egyptian Cobra: III. The otic capsule, palate, temporal bones, lower jaw and hyoid apparatus of the adult osteocranium. *Acta Zoologica* 51, 31–42.

Kardong, K.V., Weinstein, S., Smith, T., 2009. Reptile venom glands: Form, function, and future. *Handbook of venoms and toxins of reptiles*, pp. 65–91.

Kochva, E., Tonsing, L., Louw, A.I., Liebenberg, N.V., Visser, L., 1982. Biosynthesis, secretion and in vivo isotopic labelling of venom of the Egyptian cobra, *Naja haje annulifera. Toxicon* 20, 615–635.

Mackessy, S.P., 2022. Venom production and secretion in reptiles. *The Journal of Experimental Biology* 225, 66.

Rosenberg, H.I., 1967. Histology, histochemistry, and emptying mechanism of the venom glands of some elapid snakes. *Journal of Morphology* 123, 133–155.

Triep, M., Hess, D., Chaves, H., Brücker, C., Balmert, A., Westhoff, G., Bleckmann, H., 2013. 3D flow in the venom channel of a spitting cobra: Do the ridges in the fangs act as fluid guide vanes? *PLoS One* 8, e61548.

Wüster, W., Thorpe, R.S., 1992. Dentitional phenomena in cobras revisited: Spitting and fang structure in the Asiatic species of *Naja* (Serpentes: Elapidae). *Herpetologica* 48, 424–434.

Young, B.A., Boetig, M., Westhoff, G., 2009. Functional bases of the spatial dispersal of venom during cobra "spitting". *Physiological and Biochemical Zoology: PBZ* 82, 80–89.

Young, B.A., Dunlap, K., Koenig, K., Singer, M., 2004. The buccal buckle: The functional morphology of venom spitting in cobras. *The Journal of Experimental Biology* 207, 3483–3494.

4 Geographic Distribution, Habitat, Behavior, Ecology, and Reproduction of Asian Cobras

4.1 INTRODUCTION

Cobras mainly live in African and Asian countries. A total of 11 species of Cobra (genus *Naja*) are found in the Asian continent. All these species belong to the Elapidae family of snakes and are responsible for many envenoming cases in Asian countries. The following sub-sections describe these Cobra species' geographical distribution, habitat, ecology, and behavior. Furthermore, since Cobra species have similar sexual dimorphism and mode of reproduction, they are described collectively in 4.13.

4.2 *NAJA ATRA*

Scientific name: *Naja atra* (Cantor, 1842)
Common name: Chinese Cobra, Taiwan Cobra

4.2.1 Geographical Distribution

The Chinese Cobra (*Naja atra*) mainly inhabits southern China (south of the Yangtze River: Guizhou, Guangdong, Guangxi, Fujian, Jiangxi, Hainan, Southern Anhui, Hunan, Western Hubei, Taiwan, Zhejiang), Hong Kong Special Administrative Region (China), northern Vietnam (local area named Chusan) and Laos (Figure 4.1) (Cantor, 1842; Wüster and Thorpe, 1991). It is one of the most widespread venomous snakes in China and Taiwan and is responsible for most snakebite accidents (https://reptile-database.reptarium.cz/species?genus=Naja&species=atra). The World Health Organization recognizes this species as a medically important Category I snake.

4.2.2 General Description

Morphologically, the skin color of the Chinese Cobra is black/brown, but in rare cases, it can also be found in brown or golden color. This species averages 1.2 to 1.5 m

DOI: 10.1201/9781003598121-4

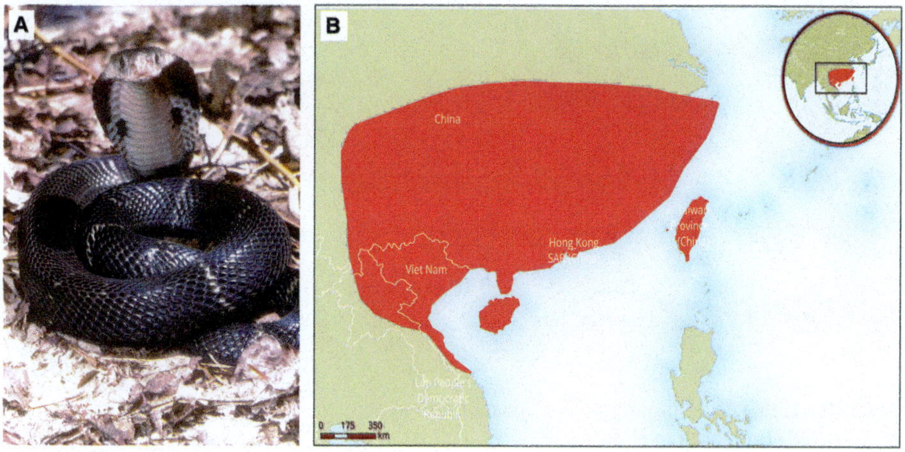

FIGURE 4.1 (A) Photograph of a *N. atra* specimen (B) Geographic distribution of *N. atra* (Longbottom et al., 2018).

(Photograph credit: Mark O' Shea.)

(3.9 to 4.9 ft) long but can grow up to 2 m (6.6 ft), although rare. At times, identification of this species can be confusing, especially if it does not present its hood, in which it can resemble a Chinese Cobra with light vertical patches running through the middle part of the body and over the entire tail. A distinctive morphological feature easily seen on its body is a white or light brown ring around its throat, partially covering its neck just behind its head. One can easily see that when the Chinese Cobra wakes up, its hood turns into a brilliant ring shape (https://eol.org/pages/458634/articles; https://www.hongkongsnakeid.com/chinese-Cobra).

4.2.3 Ecology and Habits

The Chinese Cobra generally lives in croplands, grasslands, shrublands, and mangroves. This snake has adapted to thrive in forests, open fields, and heavily populated areas. It can be seen in paddy fields and various hill forests, but it usually avoids dense forest areas. Chinese Cobras inhabit tropical forests and river banks during the rainy season. The primary population density is in human settlements, secondary forests, and agricultural paddy fields. This snake can live at ground level and in hills and mountains more than 2,000 meters (6,600 ft) high. It usually hides under logs, leaves, and rocks.

4.3 *NAJA KAOUTHIA*

Scientific name: *Naja kaouthia* (Lesson, 1831)
Common name: Monocled Cobra

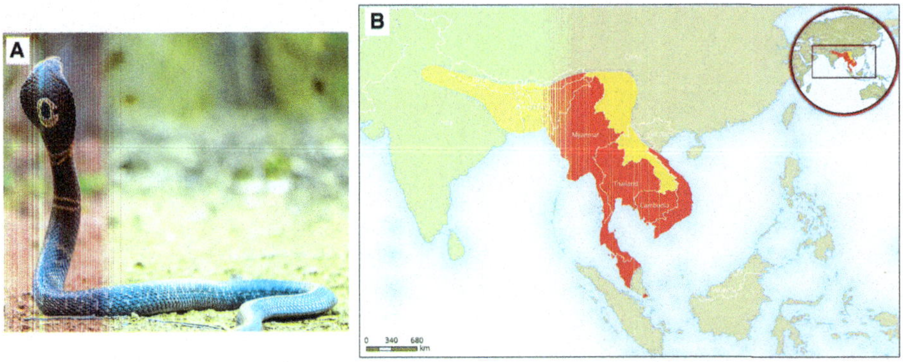

FIGURE 4.2 (A) Photograph of a *N. kaouthia* specimen, (B) Geographic distribution of *N. kaouthia* (Longbottom et al., 2018).

(Photograph credit: Subhadeep Dey.)

4.3.1 Geographical Distribution

The Monocled Cobra (*Naja kaouthia*) is distributed in Bangladesh, Myanmar, north-east India (Assam, Bihar, Mizoram, Nagaland, Sikkim, Tripura), Orissa, Uttar Pradesh, West Bengal, Nepal, Bhutan, Cambodia, Laos, North Malaysia, South China (Yunnan), Thailand, and Vietnam (Figure 4.2) (Lesson, 1831) (https://reptile-database.reptarium.cz/species?genus=Naja&species=kaouthia). This Cobra species is highly venomous and is responsible for many deaths and morbidity in the countries listed above. By the World Health Organization, *N. kaouthia* has been recognized as a medically important Category I snake.

4.3.2 General Description

The Monocled Cobra can be identified with a distinctive O-shaped hood mark. Body skin color can range from brown to black with or without clearly defined cross bands on the dorsal surface. Due to similar skin color, this species can sometimes be confused with the rat snake (*Pitus mucosa*) and the checkered keelback (*Xenocrophis piscator*) (Wüster and Thorpe, 1991). An adult Monocled Cobra can grow up to a length of 1.35 to 1.5 m (4.4 to 4.9 ft) and, in rare cases, up to 2.3 m (7.5 ft) (https://indiabiodiversity.org/species/show/227801).

4.3.3 Ecology and Habits

Monocled Cobras (*N. kaouthia*) live primarily in non-aquatic environments and have excellent climbing and swimming abilities. These can be seen in the evening and often at night. They are adaptable to habitats in both natural and artificial environments. They can be seen in primary and secondary woodlands, agricultural forests, and rural and urban localities. This serpent hides behind the dumping of wooden logs, brick piles, caves, rat holes, cracks, etc. If upset or provoked, they may raise their hood and spit forcefully in the victim's face (www.indiabiodiversity.org).

4.4 *NAJA MANDALAYENSIS*

Scientific name: *Naja mandalayensis* (Slowinski & Wüster 2000)
Common name: Mandalay Spitting Cobra, Burmese Spitting Cobra

4.4.1 Geographical Distribution

Mandalay spitting Cobra (*Naja mandalayensis*) is commonly found in Myanmar (formerly known as Burma), including the Mandalay, Magwe, central dry zone, and Sagaing Divisions (Figure 4.3) (Slowinski and Wüster, 2000) (https://reptile-database.reptarium.cz/species?genus=Naja&species=mandalayensis). It has been recognized as a Category I medically important snake by the World Health Organization.

4.4.2 General Description

This snake is muscular and of average length. The standard length of an adult snake can range from 1 to 1.2 m (3.3 to 3.9 ft), and its body can grow up to 1.4 m (4.6 ft). The head is asymmetrical, slightly modified by the neck, and has a short, flattened snout and large nasal openings. The body is flattened dorsally and sub-cylindrical posteriorly. The hindquarters are plain and highly oblique. They have a pair of fixed upper fangs, which are modified to "hiss" venom with force, similar to other spitting Cobra species (https://inaturalist.ca/taxa/30491-Naja-mandalayensis).

4.4.3 Ecology and Habits

This snake is found primarily in the central desert region of Myanmar, a place where rainfall is less than 1,000 mm (39 in) annually. The area is covered with acacia and Dipterocarpaceae savanna, which are continuously used as agricultural land. This Cobra is found in dehydrated and acacia forests (Slowinski and Wüster, 2000).

FIGURE 4.3 (A) Photograph of a *N. mandalayensis* specimen, (B) Geographic distribution of *N. mandalayensis* (Longbottom et al., 2018).

(Photograph credit: Joseph Bruno Slowinski.)

They spit exactly like the Indochinese spitting Cobra (*N. siamensis*), although herpetologists view this behavior as unscientific. This Cobra species is very alert and immediately deploys a defensive mechanism by raising and spreading a large hood when threatened. It spits when provoked or agitated. It can throw its poison to a distance of at least 2 meters (6.6 ft). Additionally, this snake makes repeated hissing sounds and tries to bite when threatened (Slowinski and Wüster, 2000).

4.5 *NAJA NAJA*

Scientific name: *Naja naja* (Linnaeus, 1758)
Common name: Indian Spectacled Cobra

4.5.1 Geographical Distribution

Indian Spectacled Cobra (*Naja naja*) is distributed throughout the Indian subcontinent, i.e., India, Pakistan, Sri Lanka, Nepal, and Bangladesh (Figure 4.4) (Linnaeus, 1758; Wüster and Thorpe, 1991). Nevertheless, this species is less abundant in northeastern India. The World Health Organization has recognized this Cobra species as a Category I medically important snake.

4.5.2 General Description

The Indian Spectacled Cobra is a highly revered snake in Indian culture, and the spectacle-like mark on its hood is believed to be the footprint of Lord Sri Vishnu (Figure 4.4). The hood is the most distinguishing feature of this snake, which is why people are attracted to and afraid of it. When the Indian Cobra senses danger, it draws the ribs near the neck in front of its body and forms a hood. This reaction is its defense mechanism. However, if an enemy or human does not heed this warning, the snake is helpless and must bite.

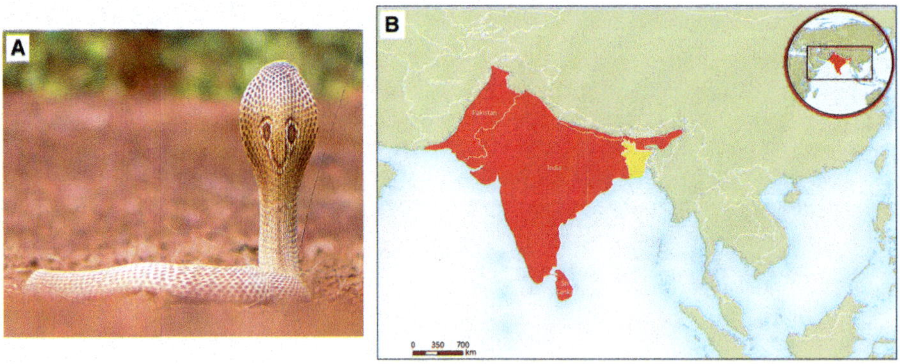

FIGURE 4.4 (A) Photograph of a *N. naja* specimen, (B) Geographic distribution of *N. naja* (Longbottom et al., 2018).

(Photograph credit: Subhadeep Dey.)

The head of the Indian Cobra is oval, flat, and slightly separated from the neck. Indian Cobras have a perfect sense of smell, and their night vision is very sharp, which dramatically helps them catch their prey.

The Indian Cobra uses its sharp, poisonous teeth (fangs) to inject venom into its prey. The arrangement of the teeth of the Indian Cobra is known as the petroglyph type. The Indian Cobra has two hollow, small, sharp teeth in the upper jaw in front of the mouth connected to the venom gland.

The average length of an Indian Cobra is about 1.0 m (3.3 to 3.5 ft); sometimes, 1.5 m (5 feet) long snakes are also seen. Indian Cobra has black eyes, a wide neck, and a body covered with smooth scales. Interestingly, there is an exciting variation in the skin color of the Indian Cobra snake in different regions of India, such as black, dark brown, brown, red, and light yellow over yellow (https://animaldiversity.org/accounts/Naja_naja/).

4.5.3 Ecology and Habits

The Indian Cobra generally prefers to live in plains, swamps, dense or open forests, and near water bodies. They can also live in mounds and tree cavities. However, densely populated urban areas with high human populations are not the preferred habitat for this snake. Old houses and temples, where human traffic is less, are preferred shelter sites for this species of Cobra. Due to the abundance of rats, the Indian Cobra also prefers to live near agricultural areas, where it can quickly get its favorite food (rats) (https://animaldiversity.org/accounts/Naja_naja/). The Indian Cobra eats rodents, lizards, and frogs.

4.6 *Naja oxiana*

Scientific name: *Naja oxiana* (von Eichwald, 1831)
Common name: Central Asian Cobra, Caspian Cobra, Ladle snake, Oxus Cobra, or Russian Cobra

4.6.1 Geographical Distribution

The Central Asian Cobra (*Naja oxiana*) is found in Afghanistan, north-east Iran, north-west India (specifically in Jammu and Kashmir), Pakistan, Kyrgyzstan, south-west Tajikistan, Turkmenistan, Uzbekistan, and Myanmar (Burma) (Figure 4.5) (von Eichwald, 1831) (https://reptile-database.reptarium.cz/species?genus=Naja&species=oxiana). For many years, it was considered a subspecies of the Indian spectacled Cobra until its genetic description revealed that it is a separate species (Wüster and Thorpe, 1991). By the World Health Organization, *N. oxiana* is recognized as a medically important Category I snake.

4.6.2 General Description

Naja oxiana is medium in length and is a thick-bodied Cobra species with long cervical scales, which can expand to form a hood. Anteriorly, the body is dorsal-ventrally

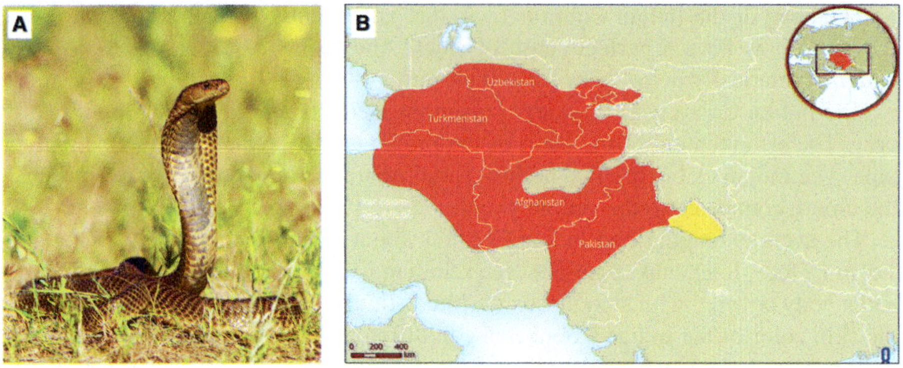

FIGURE 4.5 (A) Photograph of a *N. oxiana* specimen, (B) Geographic distribution of *N. oxiana* (Longbottom et al., 2018).

(Photograph credit: Ali Khani.)

depressed, and posteriorly, it is sub-cylindrical. The average size of this snake is about 1 to 1.4 meters (3.3 to 4.6 ft). Like other Cobra species, *Naja oxiana* can sometimes grow up to 1.7 m (5.6 ft) long (Nasoori et al., 2016). The skin color of an adult snake is entirely faint brown or yellow, while a few younger snakes have dark-colored ventral bands. There is no hood mark or lateral patches on the throat on this specimen of Cobra (https://inaturalist.ca/taxa/30491-Naja-mandalayensis).

4.6.3 Ecology and Habits

Naja oxiana is often observed in desert, dry, rocky mountain areas up to about 3,000 m (9,800 ft) above ground level. This snake is also a western species of Asian Cobra. This snake tries to stay away from humans completely, except when it faces any danger or disturbance (http://www.toxinology.com/fusebox.cfm?fuseaction=main.snakes.display&id=SN0042). The Central Asian Cobra derives its nutrition from small mammals, amphibians, and occasionally fish, birds, eggs, and other snakes (Nasoori et al., 2016).

4.7 *Naja philippinensis*

Scientific name: *Naja philippinensis* (Taylor, 1922)
Common name: Philippine Cobra

4.7.1 Geographical Distribution

Naja philippinensis, commonly known as the Philippine Cobra, is available mainly in the region of the Philippines and local areas of Pangasinan Province and Luzon (Figure 4.6) (Taylor, 1922; Wüster and Thorpe, 1991) (https://reptile-database.reptarium.cz/species?genus=Naja&species=philippinensis). This species is also recognized as a Category I medically important snake by the World Health Organization.

FIGURE 4.6 (A) Photograph of a *N. philippinensis* specimen, (B) Geographic distribution of *N. philippinensis* (Longbottom et al., 2018).

(Photograph credit: Wolfgang Wüster.)

4.7.2 General Description

This snake is thick and of medium length. The Philippine Cobra has long cervical ribs for extension, which help form a hood when the snake is in danger. The standard length of this snake is 3.3 feet, about 1.0 m, and can grow up to 5.2 feet (1.6 m). Adult snakes are light to medium brown, and younger snakes are dark brown. In Tagalog, this snake is called "ulupong," and in Locano, this snake is called "karasen".

4.7.3 Ecology and Habits

This Philippine Cobra lives in ground surface areas, forest areas, dense forests, grasslands, croplands, and human habitations. Specifically, this elapid prefers to live in aquatic environments such as ponds, rivers, and large pools. It mainly eats small mammals (mice), snakes, and frogs.

4.8 *NAJA SAGITTIFERA*

Scientific name: *Naja sagittifera* (Wall, 1913)
Common name: Andaman Cobra

4.8.1 Geographical Distribution

The Andaman Cobra (*N. sagittifera*) is found only in the Andaman Islands of India (Figure 4.7) (Wall, 1913) (https://reptile-database.reptarium.cz/species?genus=Naja&species=sagittifera). It is recognized as a medically necessary Category II snake by the World Health Organization.

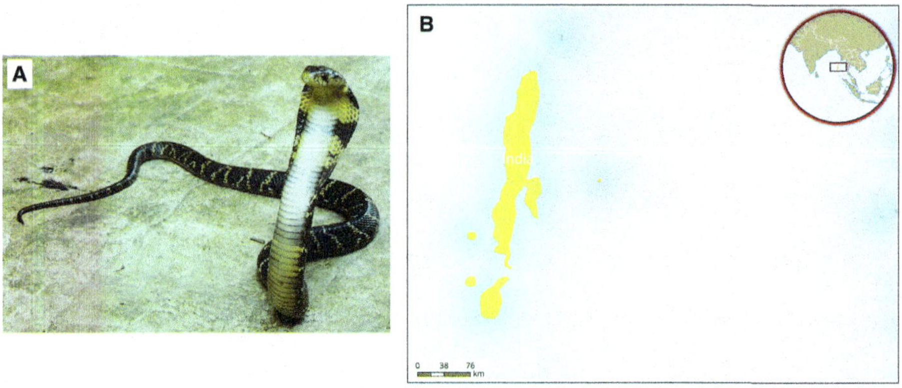

FIGURE 4.7 (A) Photograph of a *N. sagittifera* specimen, (B) Geographic distribution of *N. sagittifera* (Longbottom et al., 2018).

(Photograph credit: Ram Vikas.)

4.8.2 General Description

This Cobra is medium to large in length and heavy in weight, with long cervical ribs that help form a hood when threatened. This species of Cobra can even spit its venom into the eyes of an injured person. The body of this Cobra species is flattened dorsally and sub-cylindrically from behind. The standard length of this Cobra is about 3.0 ft (0.9 m), but from time to time, they can grow up to 4.9 ft (1.5 m) long, although the longest specimens of 5.9 ft (1.8 m) have also been observed (https://reptile-database.reptarium.cz/species?genus=Naja&species=sagittifera).

4.8.3 Ecology and Habits

The ecology and behavior of this species of Cobra living in the Andaman Islands have yet to be discovered (Attarde et al., 2021).

4.9 *NAJA SAMARENSIS*

Scientific name: *Naja samarensis* (Peters, 1861)
Common name: Samar Cobra

4.9.1 Geographical Distribution

This species of Cobra is also called "Peters' Cobra", "Southern Philippine Cobra", or "Visayan Cobra" in the local region. The Samar Cobra is a highly venomous species that inhabits the Visayas and Mindanao islands of the Philippines (Figure 4.8) (Peters, 1861; Wüster and Thorpe, 1991). This species is listed as a medically important Category I snake by the World Health Organization.

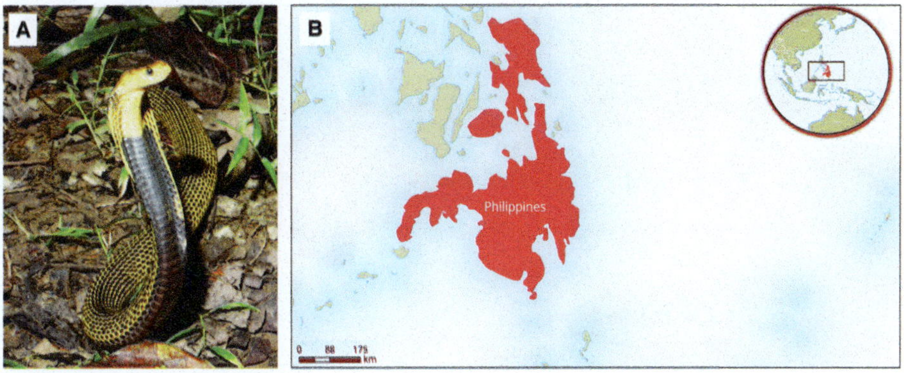

FIGURE 4.8 (A) Photograph of a *N. samarensis* specimen, (B) Geographic distribution of *N. samarensis* (Longbottom et al., 2018).

(Photograph credit: Wolfgang Wüster.)

4.9.2 General Description

The Samar Cobra can reach up to 1.4 m (4.6 ft) in length. Their body color can vary from black and yellow to green. Like the Philippine Cobra (*N. philippinensis*), this species also feeds on small rats and mice for nutrition. They may also eat frogs and other small reptiles. They feed mainly from agricultural lands and terrestrial colonies (https://reptile-database.reptarium.cz/species?genus=Naja&species=samarensis).

4.9.3 Ecology and Habits

They can live in a wide range, from mountainous forests to equatorial regions. They prefer to live in places near human habitation. They can also obtain nutrition by eating small rodents and reptiles. Humans frequently encounter this Cobra species as their primary food source is agricultural fields and human areas (https://reptile-database.reptarium.cz/species?genus=Naja&species=samarensis). Their habitat ranges from tropical dense forests to croplands, pineapple orchards, coconut plantations, and rural areas (Paghubasan et al., 2023).

4.10 *NAJA SIAMENSIS*

Scientific name: *Naja siamensis* (Laurenti, 1768)
Common name: Indo-Chinese Spitting Cobra, Siamese Spitting Cobra, Thai Spitting Cobra, Black and White Spitting Cobra

4.10.1 Geographical Distribution

The Indo-Chinese Spitting Cobra is found in Myanmar, Thailand, Cambodia, South Vietnam, and Laos (Figure 4.9) (Laurenti, 1768; Wüster and Thorpe, 1991) (https://reptile-database.reptarium.cz/species?genus=Naja&species=siamensis). This species is also listed as a Category I medically important snake by the World Health Organization.

FIGURE 4.9 (A) Photograph of a *N. siamensis* specimen, (B). Geographic distribution of *N. siamensis* (Longbottom et al., 2018).

(Photograph credit: Thai National Parks, Thailand.)

4.10.2 General Description

This Cobra is medium in size and thinner than other Cobras. They are mainly brown, gray, and black with white dots or stripes. The white part covers most of the snake's body. The spectacle-shaped hood mark on the body of adult snakes becomes irregular or absent. An adult Indo-Chinese spitting Cobra has an average length of 0.9 to 1.2 m (3 to 4 ft) and can grow to 1.6 m (5 ft). This species' habitat and morphological characteristics are similar to those of the Monocled Cobra (*N. kaouthia*). (https://www.thainationalparks.com/species/naja-siamensis).

4.10.3 Ecology and Habits

This snake inhabits many areas, such as mountains, plains, and croplands. It is also seen in dense forests and human habitations. Generally, with a high human population, the population of rats and small rodents becomes high. This snake is mainly a nocturnal species of Cobra. The snake is in danger during the day, so it looks for some dark place nearby (https://www.thainationalparks.com/species/naja-siamensis).

4.11 *NAJA SPUTATRIX*

Scientific name: *Naja sputatrix* (BOIE, 1827)
Common name: Javan Spitting Cobra, Indonesian Cobra, Komodo Spitting Cobra

4.11.1 Geographical Distribution

The Javan Spitting Cobra is commonly found in Indonesian regions like Java, Lombok, Bali, Flores, Sumbawa, Komodo, Lonblen, Alor, and Sulawesi (Figure 4.10) (Boie, 1827) (https://reptile-database.reptarium.cz/species?genus=Naja&species=sputatrix). This species is considered a Category I medically important snake by the World Health Organization.

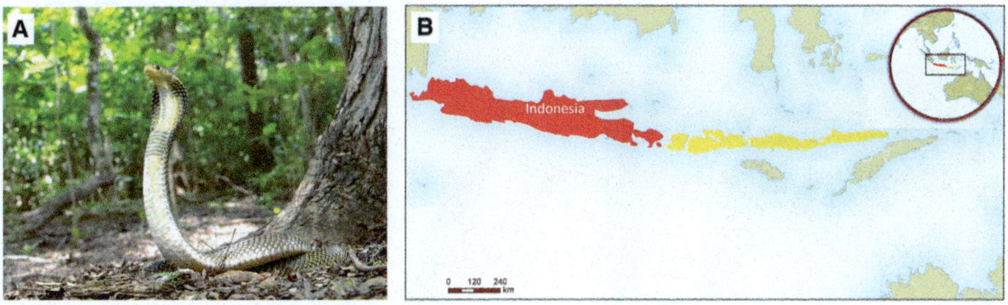

FIGURE 4.10 (A) Photograph of a *N. sputatrix* specimen, (B) Geographic distribution of *N. sputatrix* (Longbottom et al., 2018).

(Photograph credit: Halvard Aas Midtun.)

4.11.2 General Description

The Java spitting Cobra is medium to large and has long cervical ribs, which help form a hood when threatened. The snake's body is flat dorsally and cylindrical from behind. The standard length of an adult snake is 4.3 feet (1.3 m), but they can grow to about 6.1 feet (1.85 m). The head is concise and slightly separated from the neck portion and has a short-curved snout with a significant nasal passage. Its eyes are of average size with round pupils. Dorsal scales are flat and very oblique (https://uk.inaturalist.org/taxa/30482-Naja-sputatrix). Adult snakes are usually yellow, brown, or black, although smaller snakes often have lateral dots and bands around their necks. This species does not have hood markings, but it is mainly chevron-shaped if there is a marking (Wüster and Thorpe, 1991).

4.11.3 Ecology and Habits

The Javan spitting Cobra species is usually seen in tropical and wetland forests, but this Cobra species also lives in islands, mountainous areas, and dry forests. It is found on Komodo Island, a dry savanna (transient to monsoon forest) (https://uk.inaturalist.org/taxa/30482-Naja-sputatrix).

4.12 *Naja sumatrana*

Scientific name: *Naja sumatrana* (Müller, 1887)
Common name: Sumatran Spitting Cobra, Equatorial Spitting Cobra

4.12.1 Geographical Distribution

The Sumatran spitting Cobra (*Naja sumatrana*) is distributed in Malaysia (East Malaysia, Peninsula, Pulau Tioman, Pulau Pangkor), Indonesia, Singapore, Thailand (Bangka, Belitung, Borneo, Sumatra), and the Philippines (Palawan: Calamian Islands, etc.) (Figure 4.11) (Müller, 1887; Wüster and Thorpe, 1991) (https://reptile-database.reptarium.cz/species?genus=Naja&species=sumatrana). This species is recognized as a Category I medically important snake by the World Health Organization.

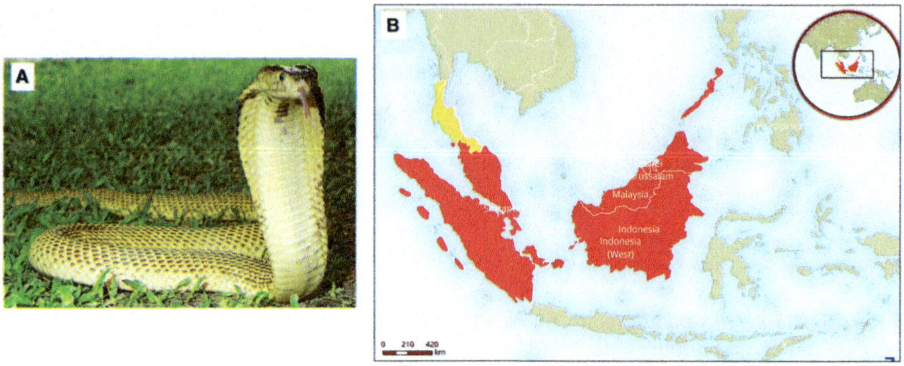

FIGURE 4.11 (A) Photograph of a *N. sumatrana* specimen, (B) Geographic distribution of *N. sumatrana* (Longbottom et al., 2018).

(Photograph credit: David Williams.)

4.12.2 General Description

This Cobra species has a medium length (0.9 to 1.2 m), but can grow up to 1.5 m. The body is flattened dorsally and sub-cylindrical posteriorly. The head of this Cobra species is abbreviated and separate from the neck. It has a short, flattened snout with a giant nostril. The eyes are of average size with round pupils. The posterior portion is plain and primarily oblique. There is no trace of hood in this species of Cobra. Depending on the geographical region, the body color of the snake species can vary from yellow to black (https://www.thainationalparks.com/species/naja-sumatrana).

4.12.3 Ecology and Habits

This species of Cobra can be seen almost in mountainous areas up to 1500 m above ground level and primary and secondary tropical regions of forests, such as dense forest areas. However, it has also been observed in urban areas such as gardens and parks where there may be human conflict. The Sumatran spitting Cobra is a terrestrial and non-nocturnal snake that feeds on rodents, frogs, and other small reptiles and mammals (https://www.thainationalparks.com/species/naja-sumatrana).

4.13 SPITTING BEHAVIOR IN ASIAN COBRAS

The distinctive spitting phenomenon, particularly in the *Naja* species, is exclusively a defensive behavior. Apart from the African Spitting Cobras, venom spitting is primarily documented for seven Asian Cobra species, including *N. atra, N. sumatrana, N. siamensis, N. sputatrix, N. philippinensis, N. mandalayensis* and *N. samarensis*. In addition, a few studies have also reported venom spitting by certain *N. kaouthia* specimens, but the phenomenon is infrequent (Santra and Wüster, 2017). In particular, some species of Cobra can eject venom as a jet spray up to a distance of 2 to 3 meters (Slowinski and Wüster, 2000). This unique spitting behavior of the Asian

Cobra species mentioned above is thought to be a projectile defense mechanism, and humans are usually affected by accident. However, it has also been noted that if spitting species are threatened, they may target the face and eyes of the attacker and spray venom as a pressurized jet (Westhoff et al., 2005). Furthermore, they have evolved to measure the distance and size of the opponent and accordingly adjust the dose of venom to be expelled through the fangs (Berthé et al., 2009).

Spitting behavior has been proposed to have evolved in parallel from their non-spitting ancestors in at least three independent lineages, once in the African Cobra, second in the Asian Cobra, and third in the rinkhals (Kazandjian et al., 2021). It is important to note that spitting Cobras have undergone subtle morphological adaptations that allow them to activate this strange venom delivery mechanism (see Chapter 3 for more details). Additionally, spitting Cobras have behavioral changes, including rapid undulating head movements that help them accurately track prey/prey movement and optimize their venom-spraying machinery (Avella et al., 2021; Berthé et al., 2009). Apart from this, some species have also demonstrated variations in how venom is secreted. For example, while most Asian spitting Cobras release venom as continuous streams, in some cases, the Indo-Chinese spitting Cobra (*N. siamensis*) delivers a fine mist of venom particles (Avella et al., 2021; Rasmussen et al., 1995; Westhoff et al., 2005).

4.14 REPRODUCTION AND LIFE CYCLE OF COBRAS

Cobras living in warm climates are usually oviparous (egg-laying animals). Oviparous females lay eggs, but no embryos develop within the eggs in the mother's body. The snake's life cycle consists of three stages: egg, baby snake or young juvenile snake, and adult Cobra.

4.14.1 Egg laying

Cobras emerge from hibernation, and their breeding season is from March to April. After mating, the female snake stores the sperm in the oviduct for 1–2 months. Under favorable conditions, female snakes lay a large number of eggs. Most snake species (about 70%) reproduce by laying eggs. However, not all eggs are guaranteed to hatch. Many eggs go bad. To ensure at least some offspring survive birth, female snakes may lay many eggs, typically 3 to 100, although the exact number of eggs depends on the snake species.

The sperm of a male snake fertilizes an egg released from the oviduct or fallopian tube of the female snake. Before laying eggs, female snakes find a place to nest, for example, a shallow hole, hollow log, or other suitable place where there is usually little traffic from other animals. The eggs hatch from an opening on the underside of the female snake's tail, called the cloaca. The female Cobra usually lays about 10 to 20 (or more) fertilized eggs under rocks, in shallow burrows, rat holes, or termite mounds. The number of eggs a snake lays each time is called a clutch. Therefore, a clutch of snakes may contain several eggs. Snake eggs are much softer than bird eggs. That is, the outer cover of the eggs of other animals (like birds) is hard, but the outer cover of snake eggs is soft and hairy. The female Cobra protects the eggs until they hatch.

The eggs are moistened by atmospheric heat or the mother's body heat. The mother stays close to the clutches and keeps them warm by coiling around the clutches so that the eggs hatch at the right time. The Cobra's ovulation period is approximately two months, and the ideal temperature range for ovulation is 78 to 84 °F (25 to 29 °C).

4.14.2 Infant and Juvenile Snakes

Female snakes guard their eggs until they hatch and at least two weeks after hatching. To hatch (give birth), the juvenile snake bites the eggshell, and the eggshell cracks. The hatchlings are between 20 and 30 cm (8 and 12 in) long. Still, it is good to remember that baby Cobras can bite as fast as adults. The potency or toxicity of their venom is approximately the same as that of an adult snake. Therefore, baby Cobra should be considered venomous.

4.14.3 Adult snakes

Baby Cobras become mature or sexually mature between 2 and 4 years after hatching, and many snakes can reach sexual maturity at about two years. There is an essential difference between a young snake and an adult snake. A young snake may shed its shell up to four times a year, but an adult snake sheds it only once a year. While molting plays an essential role in the development of various animals other than insects, it does not play a significant role in the development of snakes.

4.15 CONCLUSION

In conclusion, the diverse array of Asian Cobras, with eleven recognized species, presents a complex landscape of snake diversity and ecological significance on the continent. Identified by their iconic hood and often distinct hood markings, these snakes play a significant role in the ecosystems they inhabit, preying upon various small animals. However, their preference for agricultural areas due to the easy availability of food often brings them into frequent conflict with humans. Considering the frequency and severity of envenomation, ten out of the eleven Asian Cobra species have been designated as Category I medically essential snakes by the World Health Organization. The spitting of venom, a notable behavior observed in seven Asian Cobra species, is an evolutionary adaptation driven by convergent evolution. This defensive mechanism enables the snakes to expel venom with remarkable accuracy towards potential threats, enhancing their chances of survival when feeling threatened. Cobras, being oviparous reptiles, lay clutches of 10 to 20 fertilized eggs, with a gestation period typically lasting 1 to 2 months. Upon hatching, young cobras undergo frequent skin shedding, or molting, occurring up to four times annually during the infant and juvenile stages. While adult cobras also molt, this process is less frequent, typically occurring once a year.

REFERENCES

Attarde, S., Khochare, S., Iyer, A., Dam, P., Martin, G., Sunagar, K., 2021. Venomics of the enigmatic andaman cobra (*Naja sagittifera*) and the preclinical failure of Indian antivenoms in Andaman and Nicobar Islands. *Frontiers in pharmacology* 12, 768210.

Avella, I., Barajas-Ledesma, E., Casewell, N.R., Harrison, R.A., Rowley, P.D., Crittenden, E., Wüster, W., Castiglia, R., Holland, C., van der Meijden, A., 2021. Unexpected lack of specialisation in the flow properties of spitting cobra venom. *Journal of Experimental Biology* 224, jeb229229.

Berthé, R.A., De Pury, S., Bleckmann, H., Westhoff, G., 2009. Spitting cobras adjust their venom distribution to target distance. *Journal of Comparative Physiology A* 195, 753–757.

Boie, F.V., 1827. Bemerkungen über Merrem's Versuch eines Systems der Amphibien, 1. Lieferung: Ophidier. *Isis van Oken* 20, 508–566.

Cantor, T., 1842. LIII.—General features of Chusan, with remarks on the flora and fauna of that island. *Journal of Natural History* 9, 481–493.

Kazandjian, T.D., Petras, D., Robinson, S.D., van Thiel, J., Greene, H.W., Arbuckle, K., Barlow, A., Carter, D.A., Wouters, R.M., Whiteley, G., Wagstaff, S.C., Arias, A.S., Albulescu, L.O., Plettenberg Laing, A., Hall, C., Heap, A., Penrhyn-Lowe, S., McCabe, C.V., Ainsworth, S., da Silva, R.R., Dorrestein, P.C., Richardson, M.K., Gutierrez, J.M., Calvete, J.J., Harrison, R.A., Vetter, I., Undheim, E.A.B., Wuster, W., Casewell, N.R., 2021. Convergent evolution of pain-inducing defensive venom components in spitting cobras. *Science* 371, 386–390.

Laurenti, J.N., 1768. *Specimen medicum, exhibens synopin reptilium emendatam cum experimentis circa venena et antidota reptilium Austriacorum*. Trattner.

Lesson, R., 1831. Catalogue des Reptiles qui font partie d'une Collection zoologique recueille dans l'Inde continentale ou en Afrique, et apportée en France par M. Lamare-Piquot. *Bulletin des Sciences Naturelles et de Géologie, Paris* 25, 119–123.

Linnaeus, C., 1758. *Caroli Linnaei systema naturae, Regnum Animale*. British Museum (Natural History).(Reprinted 1939).

Longbottom, J., Shearer, F.M., Devine, M., Alcoba, G., Chappuis, F., Weiss, D.J., Ray, S.E., Ray, N., Warrell, D.A., Ruiz de Castaneda, R., Williams, D.J., Hay, S.I., Pigott, D.M., 2018. Vulnerability to snakebite envenoming: a global mapping of hotspots. *Lancet* 392, 673–684.

Müller, F., 1887. Fünfter Nachtrag zum Katalog der herpetologischen Sammlung des Basler Museums. *Verhandlungen der Naturforschenden Gesellschaft in Basel* 8, 249–296.

Nasoori, A., Shahbazzadeh, D., Tsubota, T., Young, B.A., 2016. The defensive behaviour of *Naja oxiana*, with comments on the visual displays of cobras. *Herpetological Bulletin* 138, 13–17.

Paghubasan, J., Aoki, Y., Tiglao, P.J.G., Sarmiento, M.J., Tan, M.A., Sarsalijo, M.S., Aquino, G.J.B., Comandante, J.D.L., Santamaria, E.B., Takahashi, K., Smith, C., Ariyoshi, K., Agosto, L.C., Warrell, D.A., 2023. A case series of samar cobra, *Naja samarensis* Peters, 1861 (Elapidae) envenomation. *Toxicon* 223, 107008.

Peters, W., 1861. Eine zweite Übersicht (vergl. Monatsberichte 1859 p. 269) der von Hrn. F. Jagor auf Malacca, Java, Borneo und den Philippinen gesammelten und dem Kgl. zoologischen Museum übersandten Schlangen. *Monatsberichte der Königlichen Preussischen Akademie der Wissenschaften zu Berlin* 1861, 683–691.

Rasmussen, S., Young, B., Krimm, H., 1995. On the 'spitting'behaviour in cobras (Serpentes: Elapidae). *Journal of Zoology* 237, 27–35.

Santra, V., Wüster, W., 2017. Natural history notes. *Naja kaouthia* (Monocled cobra). Behavior/Spitting. *Herpetological Review* 48, 455–456.

Slowinski, J.B., Wüster, W., 2000. A new cobra (Elapidae: *Naja*) from Myanmar (Burma). *Herpetologica* 56, 257–270.

Taylor, E., 1922. The snakes of the Philippine Islands. *Manila (Bureau of Printing or Science), Monograph* 16: 1–312.

von Eichwald, C.E., 1831. Zoologia Specialis quam expositis animalibus tum vivis, tum fossilibus potissimum Rossiæ in universum, et Poloniæ in specie… edidit DE Eichwaldus.

Wall, F., 1913. A popular treatise on the common Indian Snakes. Part 20, *Naja tripudians*. *J. Bombay Nat. Hist. Sot* 22, 243–259.

Westhoff, G., Tzschätzsch, K., Bleckmann, H., 2005. The spitting behavior of two species of spitting cobras. *Journal of Comparative Physiology A* 191, 873–881.

Wüster, W., Thorpe, R., 1991. Asiatic cobras: systematics and snakebite. *Experientia* 47, 205–209.

5 A Comparison of Venom Composition of Asian Cobras

Geographical, Intra- and Inter-Species Venom Variability

5.1 SNAKE VENOM: AN EVOLUTION-DRIVEN OFFENSE-DEFENSE ARSENAL

Snake venom produced by the venom glands is a complex mixture of bioactive molecules, including proteins, peptides, and metabolites, with the former two classes being the most abundant biomolecules expressed. A series of these proteins and peptides have evolved through positive Darwinian evolution to help the snake immobilize, kill, and digest its prey (Casewell et al., 2013; Daltry et al., 1996; Fry, 2005; Ohno et al., 1997). The evolution of venom components in the venom glands has fascinated several toxinologists.

It has been proposed that the evolution of snake venom toxins is an adaptive process involving the duplication of normal body genes encoding proteins of vital regulatory pathways (non-toxic proteins) and subsequent specific expression of the duplicated genes in the venom gland. Due to further duplication events along with deletion of a few copies and alteration of others to non-functional ones (called pseudogenes), a toxin multigene family is formed, representing a characteristic feature of the 'Birth-and-Death Model' of evolution (Casewell et al., 2020; Fry et al., 2003). Moreover, gene neomorphic mutations also contribute to expanding the diversity among the members of the multigene toxin families (Fry, 2005; Ohno, 2013). Interestingly, it has been found that the newly evolved toxins retain the ancestral (non-toxin) protein scaffolds but introduce discrete modifications in functional amino acid residues that confer a myriad of new activities to the newly developed toxins (Fry, 2005; Fry et al., 2003).

The above example is detailed for the three-finger toxins (3FTx), an abundant venom component of the Asian Cobras. It has been found that most of the 3FTx family members have evolved from a common ancestor with neuro-modulatory activities. Deleting two specific ancestral cysteine residues in the parent toxin by mutational events has led to the evolution of more potent neurotoxic activities of 3FTxs,

DOI: 10.1201/9781003598121-5

specifically the neurotoxins (Koludarov et al., 2023). Notably, during evolution, the snake species' habitat and the prey's type and availability play essential roles in molding the qualitative and quantitative distribution of the venom components. Subsequently, the relative distribution of diverse classes of toxins in a particular venom dictates the pharmacological effects of the venom on its prey/victim (Gutierrez et al., 2017; Mukherjee et al., 2016).

The mechanism of action of snake venoms can be categorized into hemotoxic, neurotoxic, and cytotoxic effects (WHO, 2010). The haemostatically active venom toxins attack the components of the cardiovascular system and affect the coagulation homeostasis in the victims. On the contrary, the neurotoxic venom components primarily act on the nervous system to induce a state of acute paralysis. Cytotoxic effects are usually localized to the envenomed site, resulting in tissue necrosis. Most viperid venoms exhibit hemotoxicity, while toxins from the Elapidae family of snakes demonstrate neurotoxic manifestations. Like most other elapids, the venom of the Asian Cobras shows potent neurotoxic and cytotoxic activities in envenomed prey/victims (Bawaskar and Bawaskar, 2004; Kularatne et al., 2009; Tan and Tan, 2015). The detailed pathophysiological effects of the Asian Cobra venoms will be discussed in Chapter 6.

Identifying and quantifying venom proteome composition is crucial for comprehending the venom-induced pathophysiological effects and designing efficient antivenoms for treating snakebites. While the technical advances in mass spectrometry-based proteomics have been essential in unraveling the detailed venom composition of the Asian Cobras in the last decade (Lomonte and Calvete, 2017; Tasoulis and Isbister, 2017), a few independent studies during the pre-proteomics era aimed at investigating the biochemical and pharmacological properties of some of the Asian Cobra venoms. In particular, these studies laid the foundation for our current understanding of the pathophysiological effects of Cobra venom on envenomed victims.

5.2 BIOCHEMICAL CHARACTERIZATION OF ASIAN COBRA VENOMS

In a study by Maity and Mukherjee from Burdwan Medical College, India, the Indian Cobra (*N. naja)* venom samples from three districts of West Bengal (Purulia, Midnapur, and Burdwan), eastern India and Maharashtra, western India, were analyzed using diverse biochemical methods to highlight the intra-species geographical venom variation (Mukherjee and Maity, 1998). The venoms were subjected to fractionation by gel filtration and ion-exchange chromatography, followed by an assay of enzymatic (protease, nucleotidase, phospholipase, and esterase) activities of crude venoms and their gel filtration fractions. Results indicated various peaks and proportions of the venom samples' acidic, basic, and neutral proteins. In particular, the *N. naja* venom samples from Burdwan and Purulia demonstrated a predominance of basic phospholipase A_2 (PLA_2) enzymes that are relatively more toxic than their acidic and neutral counterparts. Furthermore, the venom samples from Burdwan and Purulia exhibited superior proteolytic and PLA_2 activities compared to those from Midnapur and western India. Concurrently, under *in vivo* conditions, the former two venom samples showed significantly higher hemolysis, edema-inducing activity, myotoxicity, and lethality than the *N. naja* venoms from Midnapur and western India.

A similar study by Shashidharamurthy et al. (2002) investigated the variations in biochemical properties of *N. naja* venom samples from eastern, western, and southern India. Interestingly, the venom sample from East India was characterized by low molecular mass (<20 kDa) proteins. In comparison, the venom samples from western and southern regions contained variable amounts of mid- and high molecular mass (>30 kDa) proteins and low molecular mass venom toxins as suggested by SDS-PAGE analyses. An assay of enzymatic activities revealed that eastern India *N. naja* venom has superior hyaluronidase and indirect hemolytic activities. At the same time, the venom samples from western and southern India demonstrated superior caseinolytic (protease) activity, indicating geographical variation in Indian Cobra (*N. naja*) venom, which may also result in differences in the toxicity of their venom (Shashidharamurthy et al., 2002).

In the same year, a comparative analysis of the venoms of *N. naja* and its congener, *N. kaouthia*, from eastern India was performed to explore the inter-species venom variability in Indian Cobras for the first time (Mukherjee and Maity, 2002). The authors observed several protein peaks and chromatographic elution profiles of the two Cobra venom samples on a Sephadex G-50 gel filtration column. Some disparities in enzymatic activity associated with each gel filtration fraction of the venom samples were also noted. In addition, the *N. kaouthia* venom exhibited superior caseinolytic, fibrinogenolytic, and acetylcholinesterase activities compared to the *N. naja* venom. These findings highlight the existence of inter-specific venom variations in Cobras.

In addition to these studies, which comprehensively examined the biochemical properties of Cobra venoms from India, some studies also explored some of the enzymatic activities of Asian Cobra venoms to correlate proteomic findings with the observed enzymatic properties. The results of these assays are summarized in Table 5.1.

5.3 LIMITATIONS OF TRADITIONAL SNAKE VENOM PROFILING METHODS AND A SHIFT TOWARDS HIGH-THROUGHPUT AND COMPREHENSIVE OMICS APPROACHES FOR BETTER IDENTIFICATION OF VENOM COMPONENTS

Major obstacles in treating snake envenomation are a comprehensive understanding of the complex venom arsenal and the generation of efficient antivenoms to attenuate the toxic effects of the constituent venom toxins. In particular, in-depth knowledge of the venom components and geographic, individual, and ontogenetic variability is indispensable for sorting venom samples to generate efficacious antivenoms. While the biochemical and pharmacological characterization of Asian Cobra venoms provided valuable insight into their complexity, identifying and quantifying the non-enzymatic Cobra venom toxins was a significant limitation of these traditional strategies. Integration of high-throughput Omics technologies, such as genomics, transcriptomics, and proteomics, can address these limitations and may assist in generating a comprehensive inventory of venom gland-specific toxins expressed by snake species.

TABLE 5.1
Summary of Enzyme Activities Determined in Some Asian Cobra Venoms

Crude venom	Venom source	PLA_2 (Unit/mg)[a]	Caseinolytic activity (Unit/mg)[b]	Fibrinogenolytic activity (Unit/mg)[c]	SVMP activity (Unit/mg)[d]	AChE (Unit/mg)[e]	NT (Unit/mg)[f]	LAAO (Unit/mg)[g]	PDE (Unit/mg)[hy]	Reference
N. atra	Zhejiang, China	350.5 ± 30.7*	0.93 ± 0.09	ND	ND	ND	ND	9.0 ± 0.5	ND	Shan et al. (2016)
N. kaouthia	Guangxi, southern China	506 ± 25.1*	1.4 ± 0.05	ND	ND	ND	0.14 ± 0.06	17.9 ± 0.4	ND	Xu et al. (2017)
N. naja	Nadia, eastern India	2017.5 ± 4.2	ND	10.0 ± 0.001	0.12 ± 0.01	4105.8 ± 0.26	4.2 ± 0.4	17.7 ± 0.03	1.3 ± 0.1	Dutta et al. (2017)
N. naja	Sindh, Pakistan	Present	0.21 ± 0.02	ND	ND	1.46 ± 0.10*	1.2 ± 0.03	28.0 ± 0.04	0.04 ± 0.001	Wong et al. (2018)
N. naja	Burdwan, eastern India	2140.5 ± 5.2	ND	ND	0.07 ± 0.01	1480.8 ± 0.80	6.2 ± 0.09	40.7 ± 0.05	0.4 ± 0.001	Chanda et al. (2018)
N. naja	Maharashtra, western India	1830.5 ± 7.6	ND	ND	0.12 ± 0.01	350.6 ± 0.20	11.8 ± 0.3	30.5 ± 0.08	13.1 ± 0.1	Chanda et al. (2019)
N. naja	Irula, southern India	1836.8 ± 9.1	ND	ND	0.3 ± 0.04	120.59 ± 0.02	15.6 ± 0.2	57.40 ± 0.01	6.3 ± 0.1	Chanda and Mukherjee (2020b)
N. naja	Karnataka, southern India	89.4 ± 0.9	ND	ND	0.5 ± 0.4	ND	ND	Present	ND	Attarde et al. (2021)

N. sagittifera	Port Blair, Andaman Island	106.6 ± 1.1	ND	ND	1.5 ± 0.8	ND	ND	Present	ND	Attarde et al. (2021)

ND: not determined.

[a] Unit activity is defined as the amount of protein that produces a decrease in 0.01 absorbance of egg-yolk suspension in 10 min at 740 nm (* determined against soybean lecithin and the unit is defined as a change in absorbance of 0.3 unit/min/μg crude venom).

[b] Unit activity is defined as nM of L-Tyrosine released/min/mg crude venom.

[c] Unit activity is defined as 1.0 μg of tyrosine equivalent liberated/min/ml of enzyme.

[d] Unity activity is defined as change in OD per minute (ΔA_{450} nm/min).

[e] Unit activity is defined as the amount of protein which hydrolyzed 0.01 mg of acetylcholine iodide in 1 min in 3 ml of 0.0092 M acetylcholine iodide (* the same assay was performed in 1 ml using 0.2 mg of acetylcholine iodide).

[f] Unit activity is defined as mM Pi released per min.

[g] Unit activity is defined as nM of kynurenic acid produced/min

[h] Unit activity is defined as mM of p-nitrophenol released per min. Values are mean ± SD of triplicate determinations.

5.4 THE REFERENCE GENOME OF THE INDIAN COBRA (*N. NAJA*)

To date, the genomes of over 20 snake species have been assembled and extensively studied (Ahmad et al., 2021; Castoe et al., 2013; Castoe et al., 2011; Schield et al., 2019; Shibata et al., 2018; Suryamohan et al., 2020; Vonk et al., 2013; Yin et al., 2016). Apart from these assembled genomes, low-coverage and unassembled genomic shotgun read datasets are available for 16 more snake species (Pasquesi et al., 2018). It is to be noted that the availability of a high-quality reference genome or a transcriptome of a particular species aids in the construction of a comprehensive and species-specific venom proteome during the database-dependent mass spectrometry analysis.

Indian Cobra *N. naja* represents the only Cobra species for which a complete genome assembly has been published. The near-chromosomal de novo genome assembly of Indian Spectacled Cobra was generated using the long-read and the short-read next-generation sequencing strategies (Suryamohan et al., 2020). The DNA used for the genome assembly was isolated from the liver tissues of a single adult male individual from Kerala, southern India. The long-read data was acquired on PacBio and Oxford Nanopore Technologies (ONT) platforms, while the short-read data was generated using the Illumina platform. The sequence reads thus obtained were combined to build contigs, scaffolds, and super-scaffolds. After that, the super-scaffold sequence reads were assigned to chromosomes with the help of SChrom-seq (single chromosome sequencing) data.

The *N. naja* reference genome was decoded with a size of 1.79 Gb, accounting for 95% of the genome. A significant proportion of the genome (~43%) was repetitive and shared close similarity with other squamate reptile genomes. Gene prediction and annotation of the sequence reads identified 23,248 protein-coding genes; among them, only 12,346 genes were expressed in the venom gland. Further, among the 12,346 venom gland-expressed genes, 139 genes belonging to 33 snake venom protein families were identified that included 19 3FTx genes, 8 snake venom metalloproteinase (SVMP) genes, 6 cysteine-rich secretory venom protein (CRISP) genes and 2 PLA_2 genes.

5.5 ADVANCES IN VENOM GLAND TRANSCRIPTOME PROFILING OF ASIAN COBRAS

Identifying venom toxins using mass spectrometry-based proteomics depends on using protein sequence databases to match the acquired MS/MS spectra. These databases, such as the NCBI or the UniProt, house primary sequences of venom toxins. As a result, if the sequence of any novel venom protein has not yet been submitted to the databases, the search results do not detect such toxins in the venom, even if the snake species express them. This failure to detect a venom toxin is one of the inevitable limitations of the database-dependent nature of mass spectrometry-based proteomics of non-model organisms, leading to the underrepresentation of venom diversity.

It was experimentally demonstrated in one of the studies from our laboratory that aimed at deciphering the venom proteome of the Indian saw-scaled viper (*Echis carinatus*) (Patra et al., 2017). When the MS/MS data obtained from gel filtration fractions of *E. carinatus* venom was searched against a species-specific (*E. carinatus*) database, only a few proteins (n=22) could be identified. However, when the target database was broadened to genus (*Echis*) and family (Viperidae) levels, the number of identified proteins increased significantly, and 47 and 90 proteins, respectively, could be determined from the same MS/MS data. Therefore, including homologous proteins from related snake species for database search can aid in increasing the proteome coverage; however, identifying novel venom toxins remains daunting.

Transcriptome profiling of the snake venom glands is one of the methods that can appropriately address this limitation of database-dependent mass spectrometry analysis (Brahma et al., 2015; Lomonte and Calvete, 2017; Sunagar et al., 2016). Venom gland transcriptome profiling provides the unique opportunity to construct species-specific databases against which the corresponding MS/MS data can be matched to create the venom proteomes. Subsequently, it may lead to identifying unknown venom toxins; thus, the proteome will be comprehensive, and the true diversity of snake venom proteomes can be appreciated. Notably, the comprehensiveness of transcriptome assemblies is primarily influenced by the quality of the mRNAs isolated from venom glands, and degradation of mRNAs during processing steps might result in the loss of some toxin transcripts. Therefore, combining the reference transcriptome with a public database (NCBI, UniProt) is recommended when performing an MS/MS search.

Transcriptome profiling has further implications in venom toxin quantitation. Current methods of calculating the relative abundance of venom toxins are based on the assumption that the target database is complete; therefore, proteins identified by MS/MS detection can be assigned a relative abundance value (Lomonte and Calvete, 2017). However, suppose there are novel proteins present in the venom that cannot yet be identified due to the limitations of database searching. In that case, they are not accounted for during the reconstruction of the quantitative venom proteome. Therefore, the calculated relative abundance of the identified venom toxins remains an approximation. Incorporation of transcriptome-derived protein sequences into the MS/MS search target database may resolve this ambiguity and provide more accurate quantitative values for the venom toxins.

The venom gland transcriptome profiling usually includes RNA extraction and purification from snake venom glands, cDNA library construction, sequencing using Next-generation sequencing (NGS) technology, and transcriptome assembly, annotation, and quantification (Holding et al., 2018). Much of the transcriptome profiling workflow is automated, and several commercial ready-to-use reagents take care of the proper handling, storage, and processing of the RNA/cDNA samples. Independent software modules perform transcriptome assembly and annotation of the sequenced transcripts. In addition, dedicated software calculates the transcript abundances based on the FPKM (Fragments per kilobase of exon model per million reads mapped) values.

To date, the venom gland transcriptomes of four Asian Cobra species, including *N. atra* (China), *N. kaouthia* (China, Malaysia, and Thailand), *N. naja* (Sri Lanka and India), and *N. sumatrana* (Malaysia) have been constructed (Chong et al., 2019; Suryamohan et al., 2020; Tan et al., 2017a; Wong et al., 2021; Xu et al., 2017). While the proteomes of these Cobra species were reported earlier with ~15 protein families, the transcriptomic analyses suggested the expression of ~22 to 23 toxin families in the venom glands. These data indicate that these Cobra species express certain toxin classes in their venom glands whose abundances are below the detection limit of the current proteomics workflows. Alternatively, some transcripts may not always be translated into toxins in the venom gland. Several species-specific gene families were identified in the venom gland transcriptomes that were otherwise not detected in the proteome analysis. For example, the expression of 15 *N. kaouthia*-specific protein families was reported in the *N. kaouthia* (Malaysia and Thailand origin) venom gland transcriptomes. These protein families include NP, KSPI, CTL, PDE, SVSP, 5'-NT, AChE, aminopeptidase, neprilysin, phospholipase B, waprin, insulin-like growth factor, dipeptidylpeptidase-IV, hyaluronidase and coagulation factor (Tan et al., 2017a).

Similarly, the venom gland transcriptomics of *N. sumatrana* from Malaysia identified 15 additional protein families that were not previously reported in the venom proteome (Chong et al., 2019). Nevertheless, concordant with the proteomic analyses, the venom gland transcriptomes of all four Asian Cobra species were predominated by the expression of 3FTxs (~80 to 97%), reinstating the fact that 3FTxs are one of the major toxin families in Cobra venoms.

Although the advantages of transcriptomics are apparent, only a few venom gland transcriptomes from Asian Cobras have been characterized. One of the reasons for such a slow increase in the number of transcriptomes compared to proteomes is the availability of venom glands for the isolation and purification of RNAs. The snake's venom gland had to be removed to isolate the RNA at the cost of its life. Therefore, obtaining ethical clearance for such an experiment becomes challenging in countries like India and elsewhere, where snake species are protected under the Wildlife Protection Act of 1972.

5.6 MASS SPECTROMETRY-BASED PROTEOMIC STRATEGIES FOR ASSEMBLING ASIAN COBRA VENOM PROTEOMES

While the genome and transcriptome assemblies are crucial, it should be noted that not all genes are transcribed into mRNA. Not all transcripts are translated into proteins. Therefore, the actual composition of a venom must be analyzed at the protein level. The introduction of the revolutionary technology of tandem mass spectrometry (MS/MS)-based proteomics study to snake venom research at the beginning of the 21st century brought about a paradigm shift to venom research, and soon, it proved its competency to address the questions about snake venom complexity. Gradually, with advances in venom decomplexation (pre-fractionation) methods and robust database search algorithms, snake venom proteomics evolved as an excellent tool for deciphering venom composition and correlating these findings on qualitative and quantitative distribution of venom toxins with extended pathophysiology post envenomation (Chanda and Mukherjee, 2020a; Georgieva et al., 2008; Kalita and Mukherjee, 2019; Tasoulis and Isbister, 2017).

Apart from a handful of studies that have tried to establish the venom proteome composition without any pre-fractionation of the Asian Cobra venom samples, most proteomic investigations essentially began with a decomplexation method followed by trypsin digestion and mass spectrometry analyses were performed for protein

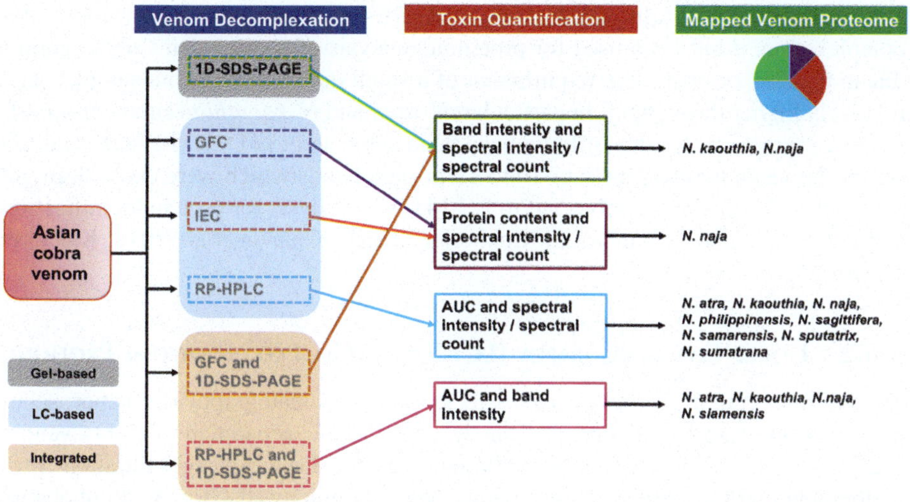

FIGURE 5.1 An overview of venom decomplexation and toxin quantification strategies used in proteomic analyses of Asian Cobra venoms.

identification. Based on the type of pre-fractionation used, the proteomics strategy can be broadly classified into gel-based, gel-free liquid chromatography, and integrated approaches (Figure 5.1). Notably, each of these strategies has its pros and cons. The choice of a particular process depends on the type of study deliverables, the amount of venom available, accessible infrastructure in the scientist's laboratory and nearby areas, availability of funds, etc.

5.6.1 Gel-Based Proteome Profiling

Gel-based strategies rely on 2D-SDS-PAGE or 1D-SDS-PAGE analysis to fractionate the venoms, followed by in-gel trypsin digestion of proteins in gel spots or gel sections and MS/MS analysis. While 2D-SDS-PAGE can separate the venom proteins based on their isoelectric point (pI) and molecular weight, 1D-SDS-PAGE fractionates the venom proteins based solely on their molecular weight. One of the very first proteomic investigations on *N. kaouthia* venom from Thailand employed 2D-SDS-PAGE for venom decomplexation followed by mass spectrometry identification of the proteins present in the gel spots that led to the identification of ~61 proteins (Kulkeaw et al., 2007).

The gel-based protein identification approach was also used for elucidating the ontogenetic variations in *N. kaouthia* venoms (Thailand) from adult and juvenile individuals (Modahl et al., 2016). Noteworthy advantages of this strategy include an extensive decomplexation of venom toxins and details on isoelectric points and molecular weights of venom toxins. However, it becomes challenging to identify low abundant venom toxins and proteins with high PI and molecular mass (Lomonte and Calvete, 2017). Furthermore, the method is labor-intensive with low throughput and is seldom used for proteome analysis of snake venoms.

On the contrary, although 1D-SDS-PAGE provides a modest venom de-complexation, it has been used for proteome analysis of some Asian Cobra venoms due to the ease of operation, requirement of a small quantity of venom samples, and relatively higher throughput. In particular, *N. naja* and *N. kaouthia* venoms from different geographical locales of India have been subjected to 1D-SDS-PAGE analysis where the venoms separated into 10 to 12 protein bands which were excised, in-gel digested with trypsin and subjected to LC-MS/MS analysis for protein identification (Chanda et al., 2019; Chanda and Mukherjee, 2020b; Chanda et al., 2018; Kakati et al., 2022).

5.6.2 Liquid Chromatography (LC)-Based (Gel-Free) Proteome Profiling

The gel-free proteomic strategy uses diverse liquid chromatography (LC) approaches to fractionate venom components. The chromatographic separation of Cobra venom components has been achieved either based on molecular weight (gel-filtration chromatography, GFC), charge (ion-exchange chromatography, IEC) or hydrophobicity (reversed-phase high-performance liquid chromatography, RP-HPLC) (Figure 5.1). The resulting protein fractions are subjected to in-solution trypsin digestion and tandem mass spectrometry analysis for protein identification. Usually, the trypsin-digested peptides are introduced into the mass spectrometer via an RP-HPLC system connected in-line with a mass spectrometer (LC-MS/MS platform) that performs the MS and MS/MS analyses. MS analysis provides an overview of the mass-to-charge (m/z) ratio of the peptide ions present in the sample. Subsequently, a few peptide ions are selected (based on abundance) for collision-induced dissociation (CID) and MS/MS measurements. Robust search engines and algorithms deconvolute the resultant MS/MS spectra and then matched them to the Elapidae database entries for protein identification.

LC-based approaches result in excellent venom de-complexation and allow extended proteome coverage. Therefore, they have been used to unravel the proteome composition of several Asian Cobra venoms. Moreover, GFC or IEC-based approaches provide the investigator with an opportunity to concurrently perform the biochemical and pharmacological characterization of the venom fractions and corroborate it with the proteomic findings (Dutta et al., 2017; Kalita et al., 2017; Mukherjee et al., 2016).

In addition, apart from serving the needs of proteome analysis, a systematic LC-based approach can also lead to the isolation and characterization of novel Cobra venom toxins. Dutta et al. (2017) separated the *N. naja* venom sample from Nadia (West Bengal), eastern India, on a HiPrep CM FF 16/10 cation-exchange (CEX) column coupled to an FPLC system that resolved the crude venom into six distinct peaks. Similarly, *N. naja* venom from western India was subjected to GFC on a HiLoad 16/600 Superdex 30 pg column coupled to an FPLC system that separated the venom into four peaks (Chanda et al., 2019). Moreover, *N. atra* venoms from eastern and western China, *N. philippinensis* venom (source not disclosed), *N. samarensis* venom from the southern Philippines, *N. sagittifera* venom from Andaman Island, *N. sputatrix* venom from Java Island, *N. sumatrana* venoms from Sumatra, Malaysia and southern Thailand, *N. naja* venom from south India, and *N. kaouthia* venom from north-eastern India were separated by RP-HPLC that resolved the venom samples into a variable number (10 to 30) of peaks (Attarde et al., 2021; Huang et al., 2015; Kakati et al., 2022; Palasuberniam et al., 2021; Tan et al., 2022; Tan et al., 2019a; Tan et al., 2017b). Subsequently, the proteins in these GFC, IEC, and RP-HPLC peaks were digested with trypsin and identified by LC-MS/MS analysis for venom proteome reconstruction.

While they are bestowed with several advantages, reconstructing the venom proteome under investigation can be pretty challenging and labor-intensive due to the peptide-centric behavior of the gel-based and LC-based approaches. The large datasets generated from the individual gel bands/fractions inevitably contain a lot of redundant peptide/protein entries spanning more than one gel band/chromatographic fraction that must be curated manually. Furthermore, since conserved peptide sequence can be present in multiple different proteins or protein isoforms, such shared peptides can lead to ambiguities in correctly assigning the identities of sample proteins, a phenomenon often known as the 'protein inference problem' (Nesvizhskii and Aebersold, 2005).

Therefore, in gel-based and LC-based peptide-centric proteomic approaches, it is always suggested that the minimum number of proteins that can account for all the peptides identified in the sample be represented. As a common practice, the number of proteins from each toxin family is determined by performing multiple sequence alignments of all the identified tryptic peptides matching the corresponding full-length sequences of each toxin, thereby identifying overlapping distinct specific peptides. After that, the occurrence of one or two overlapping specific peptides in a protein entry is set as a pre-requisite for unambiguous identification (Chanda et al., 2019; Chanda et al., 2018; Chapeaurouge et al., 2015; Dutta et al., 2017).

5.6.3 The Venomics Approach: Integrating Gel-Free and Gel-Based Strategies

Calvete and his colleagues introduced the venomics approach in which the venom is initially fractionated using RP-HPLC. After that, the proteins present in the resolved fractions are further separated by 1D-SDS-PAGE analysis. The protein bands are excised from the gel, in-gel digested with trypsin, and then the peptides are submitted to tandem mass spectrometry analysis for protein identification (Calvete et al., 2007). This integrated method has also been used to decipher the venom proteomes of several Asian Cobra species, including *N. atra* (China and Taiwan), *N. siamensis* (Thailand), *N. naja* (Sri Lanka, Pakistan, western India, and north-western India), and *N. kaouthia* (Thailand, southern China, Malaysia, and Vietnam) (Chanda et al., 2019; Laustsen et al., 2015; Liu et al., 2017; Shan et al., 2016; Sintiprungrat et al., 2016; Tan et al., 2015; Wong et al., 2018; Wong et al., 2021; Xu et al., 2017).

Although widely accepted, this approach is also associated with a few caveats, such as the difficulty eluting high molecular weight hydrophobic proteins from the C_{18} columns used in the RP-HPLC system. Furthermore, less abundant and glycosylated Cobra venom toxins weakly stained by the Coomassie brilliant blue stain may also go undetected during the SDS-PAGE analysis. Therefore, such venom proteins will likely be missed during the downstream proteome analysis (Lomonte and Calvete, 2017).

5.6.4 Quantification of Venom Toxins

While protein identification is one aspect of mapping venom proteomes, quantifying individual venom toxins or toxin classes is also crucial because the quantitative distribution of a particular type of venom toxin can dictate the pathophysiological effects of Cobra venoms. Notably, one must adopt a labeling-based approach where isotope-labeled venom toxin peptides can serve as internal standards to determine the absolute amount of Cobra venom components. However, such a strategy has potential limitations, including high cost of reagents, higher sample requirement, increased time and complexity of sample preparation, etc. (Zhu et al., 2010).

To date, absolute quantification for Asian Cobra venom proteomes has not been attempted. Instead, it is preferable to represent the relative abundance of the individual Cobra venom toxins or toxin classes. Several innovative strategies have been developed to determine the relative abundance of the Cobra venom toxins, and they can be broadly classified into the following three broad categories.

(i) The area under RP-HPLC curve @215 nm (AUC) coupled to SDS-PAGE densitometry,
(ii) SDS-PAGE densitometry coupled with label-free quantification approaches and
(iii) AUC or protein content coupled to label-free quantification approaches (Figure 5.1).

Since they rely on discrete working principles, minor variation in the relative toxin abundances calculated by these methods is always anticipated (Tasoulis and Isbister, 2017).

5.6.4.1 Area Under RP-HPLC Curve Coupled to SDS-PAGE Densitometry

This method of relative abundance calculation is predominantly followed for the venomics approach. The area under the 215 nm curve is considered a surrogate measure for calculating toxin relative abundances where the sum of all the peak areas is 100%. If a protein peak contains more than one protein, the SDS-PAGE band intensities of the constituent proteins are accounted for normalizing the relative abundance (Lomonte and Calvete, 2017). As evident from the pipeline, the method is comparatively slow and requires extensive experimental work, which impacts throughput. Moreover, the influence of the side chain composition of the aromatic amino acids on the absorbance values at 215 nm may also lead to inaccurate relative abundance calculations (Antosiewicz and Shugar, 2016). The abundance values may also be affected by proteins present in small amounts or by underestimation of glycosylated Cobra venom toxins due to conventional staining limitations (Neuhoff et al., 1985; Osset et al., 1989).

5.6.4.2 SDS-PAGE Densitometry Coupled with Label-Free Quantification Approaches

This method is employed in the gel-based proteome profiling strategy, and it relies on the SDS-PAGE band intensities to represent the relative abundance of the venom toxins. Whenever more than one protein is present in an SDS-PAGE band, the relative abundance is normalized by label-free quantification (LFQ) methods, either

MS1-based peptide peak area/intensity or MS2-based spectral counting. While the MS1-based LFQ integrates the chromatographic peak areas for an identified peptide, spectral counting assumes that peptides with greater abundances are selected for CID-MS and will yield more MS/MS spectra (Zhu et al., 2010). Integration of MS1 peak area/intensity is usually accomplished by dedicated algorithms that are integrated into the database search engines/software. While there are no specialized algorithms for the spectral counting method, it is necessary to normalize the spectral counts by number of identified peptides or protein length (Choi et al., 2008). Notably, due to the relative ease of experimentation and requirement of a small sample, the LFQ method has been widely used to calculate relative toxin abundances in diverse Asian Cobra venoms. However, this method also demands extensive manual curation of the MS data that affects the throughput.

5.6.4.3 Protein Content Coupled to Label-Free Quantification Approaches

This strategy is primarily used for LC-based approaches where the protein content of the chromatographic fractions is estimated, and the relative contribution of each fraction is assigned. If more than one protein is identified in a particular fraction, the relative abundance is again normalized by LFQ methods (Chanda et al., 2019; Dutta et al., 2017; Kalita et al., 2018b; Kalita et al., 2017; Patra et al., 2017). Unfortunately, although very straightforward, LFQ strategies are very much database-dependent due to their peptide-centric behavior, and in the absence of a comprehensive database, they can be influenced by shared peptides (peptides that are present in more than one protein entry) (Lomonte and Calvete, 2017) resulting in wrong interpretation.

It is evident that all these methods are associated with advantages and some limitations, and the calculated relative toxin abundance is always an approximation. Therefore, as a form of validation, it is highly advisable to correlate the relative abundance of the venom toxins with the SDS-PAGE band intensities of crude venom. A good corroboration between the relative SDS-PAGE band intensity of crude venom in designated mass ranges and the relative abundance of toxin families calculated by LFQ analysis has been repeatedly demonstrated in several proteomic studies from our lab as well as by other scientists which reinforces the feasibility of LFQ methods in snake venom proteome analysis (Chanda et al., 2019; Chanda and Mukherjee, 2020b; Chanda et al., 2018; Kakati et al., 2022; Kalita et al., 2018b; Kalita et al., 2017; Patra et al., 2019; Patra et al., 2017). Moreover, a recent proteomic investigation on *N. kaouthia* venom from north-eastern India showed comparable findings on relative abundances of venom toxins calculated by LFQ methods and the venomics approach (Kakati et al., 2022).

5.7 COMPOSITIONAL ANALYSIS OF ASIAN COBRA VENOMS USING PROTEOMIC PLATFORMS

The quantitative venom proteomes of all eleven species of Asian Cobras have been unraveled using mass spectrometry-based proteomics analysis. These venom proteomes have been represented by ~20 to 100 snake venom toxins distributed over more than 15 enzymatic and non-enzymatic protein families. Notably, the classification of toxin families of *N. mandalayensis* venom proteome (represented by 86 proteins) from Myanmar was done based on UniProt 'molecular function' and 'mechanism of action' identifiers rather than on snake venom protein families (Beraldo Neto et al., 2021). Furthermore, except for 3FTxs and snake venom metalloproteases (SVMP), the relative abundances of other protein families identified in *N. oxiana* venom from northern India were not well represented (Deka et al., 2023).

In another recent study from Pakistan, the venom proteome of *N. oxiana* was analyzed, representing the relative abundance of venom classes in terms of the number of identified proteins (toxins), which made comparisons unclear and was thus excluded (Manuwar et al., 2020). Apart from these two Asian Cobra species, the rest of the venom proteomes have been analyzed and are uniformly represented (Attarde et al., 2021; Chanda et al., 2019; Chanda and Mukherjee, 2020b; Chanda et al., 2018; Dutta et al., 2017; Huang et al., 2015; Kakati et al., 2022; Laustsen et al., 2015; Liu et al., 2017; Palasuberniam et al., 2021; Ryabinin et al., 2019; Shan et al., 2016; Sintiprungrat et al., 2016; Tan et al., 2022; Tan et al., 2019a; Tan et al., 2017a; Tan et al., 2015; Wong et al., 2018; Wong et al., 2021; Xu et al., 2017).

5.7.1 Distribution of Enzymatic Toxins in Diverse Asian Cobra Venoms

Enzymatic proteins catalyze specific biochemical reactions that have profound pathophysiological outcomes. However, Cobra venoms contain relatively lower amounts of enzymatic proteins than viperid venoms. The enzymatic protein families identified in diverse Asian Cobra venoms include PLA_2, SVMP, snake venom serine protease (SVSP), 5'-nucleotidase (5'-NT), phosphodiesterase (PDE), L-amino acid oxidase (LAAO), acetylcholinesterase (AChE), and cholinesterase (ChE) (Figure 5.2–5.10).

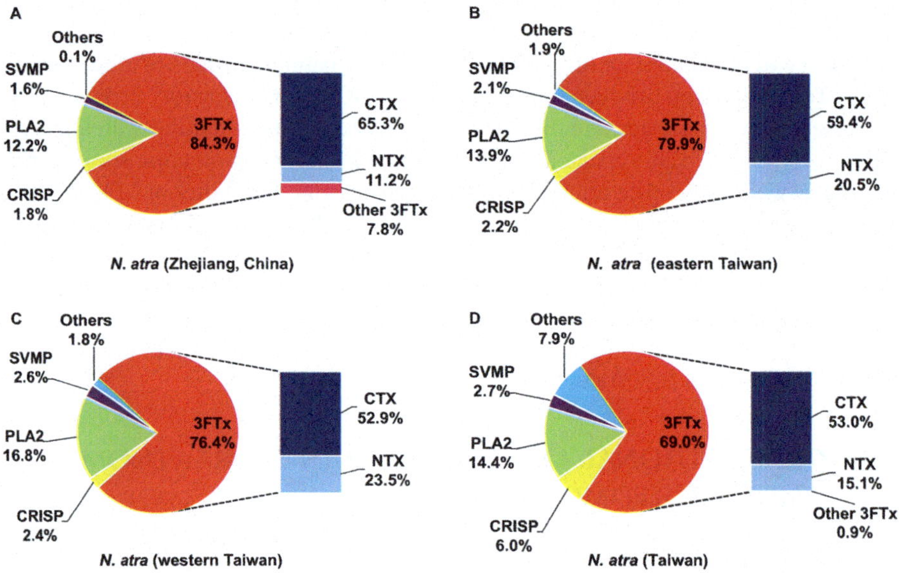

FIGURE 5.2 Venom proteome composition of *N. atra* from (A) Zhejiang, China, (B) eastern Taiwan, (C) western Taiwan, and (D) Taiwan.

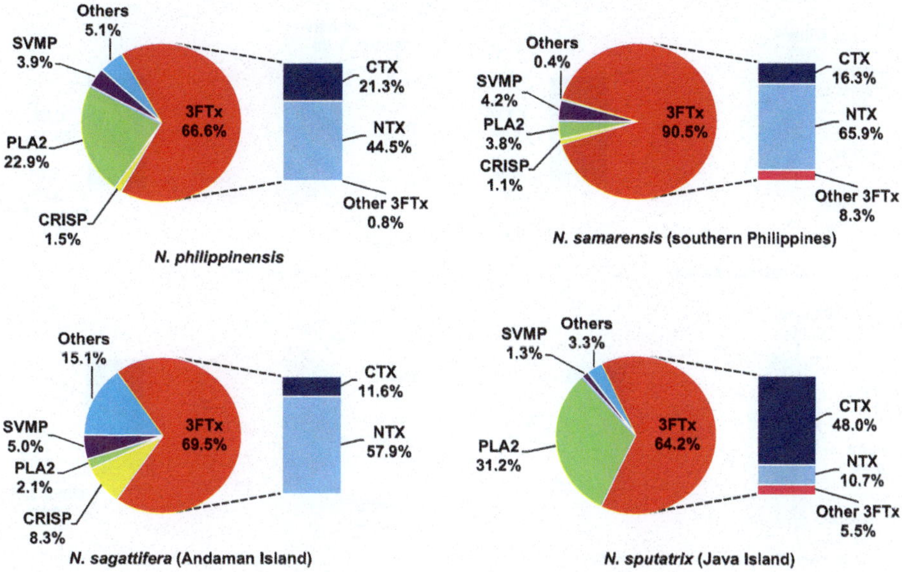

FIGURES 5.3–5.6 Venom proteome compositions of *N. philippinensis, N. samarensis* (southern Philippines), *N. sagattifera* (Andaman Island), and *N. sputatrix* (Java Island), respectively.

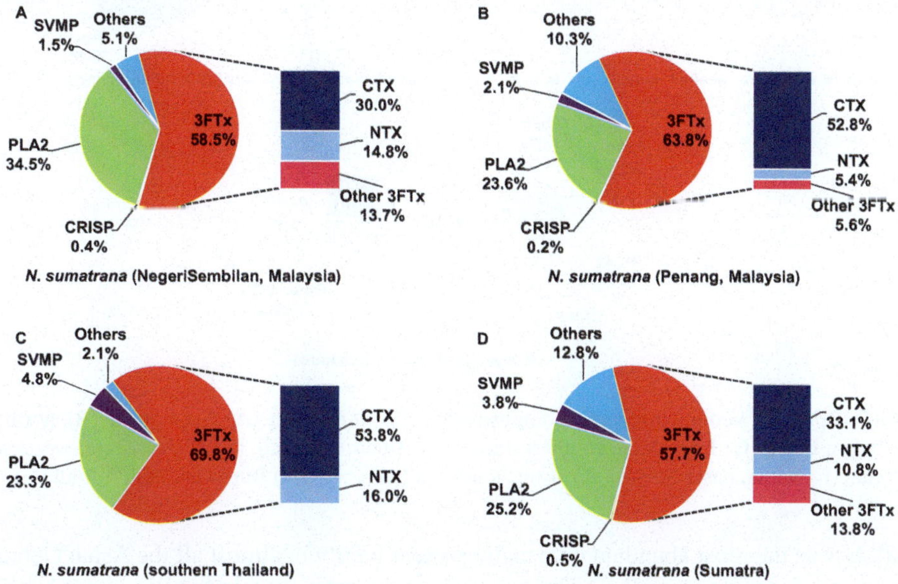

FIGURE 5.7 Venom proteome composition of *N. sumatrana* from (A) Negeri Sembilan, Malaysia, (B) Penang, Malaysia, (C) southern Thailand, and (D) Sumatra.

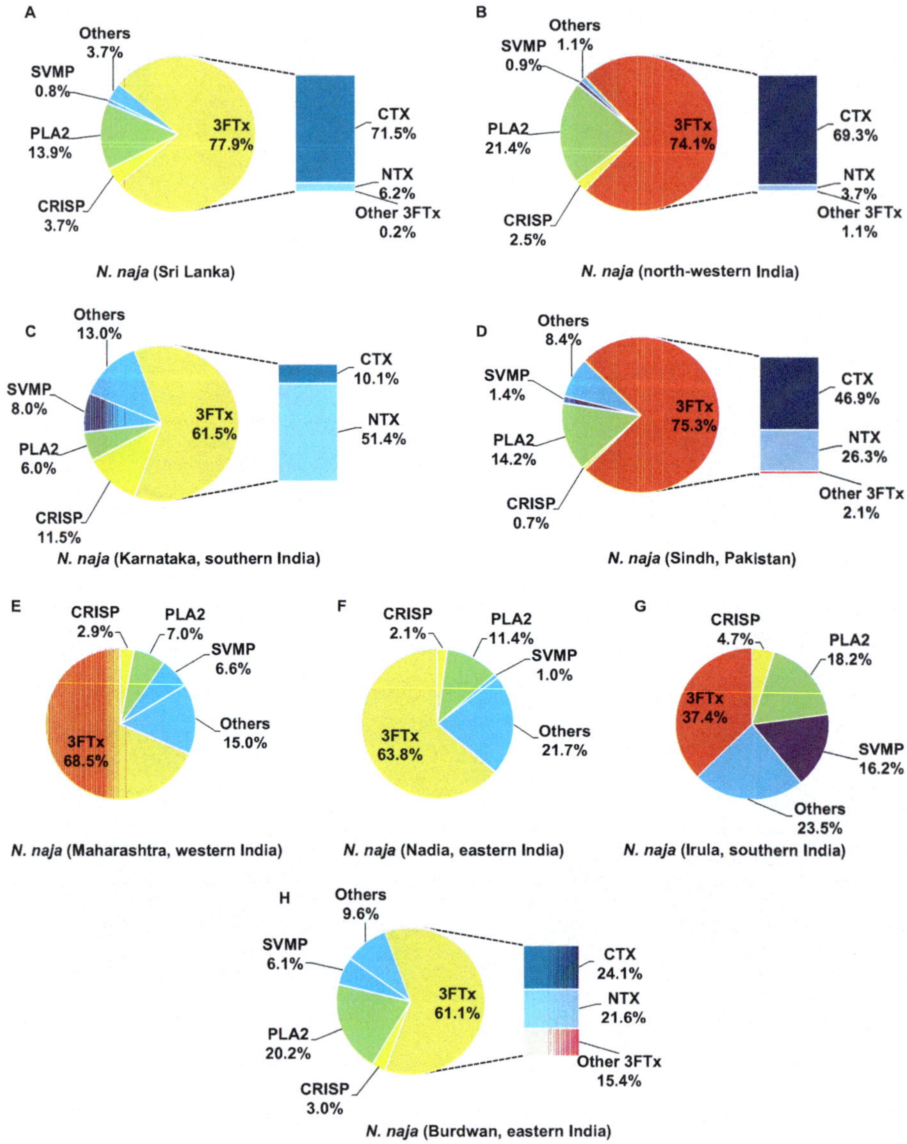

FIGURE 5.8 Venom proteome composition of *N. naja* from (A) Sri Lanka, (B) north-western India, (C) Karnataka, southern India, (D) Sindh, Pakistan, (E) Maharashtra, western India, (F) Nadia, eastern India, (G) Irula, southern India, and (H) Burdwan, eastern India.

PLA_2 was the most abundant enzymatic protein family in almost all the Asian Cobra venoms, followed by SVMP. The remaining enzymatic protein families contributed to <5% of the venom proteome in most Asian Cobra species. Details of the enzymatic proteins and relative abundances of toxin classes identified in the Asian Cobra venom proteomes are summarized in Table 5.2.

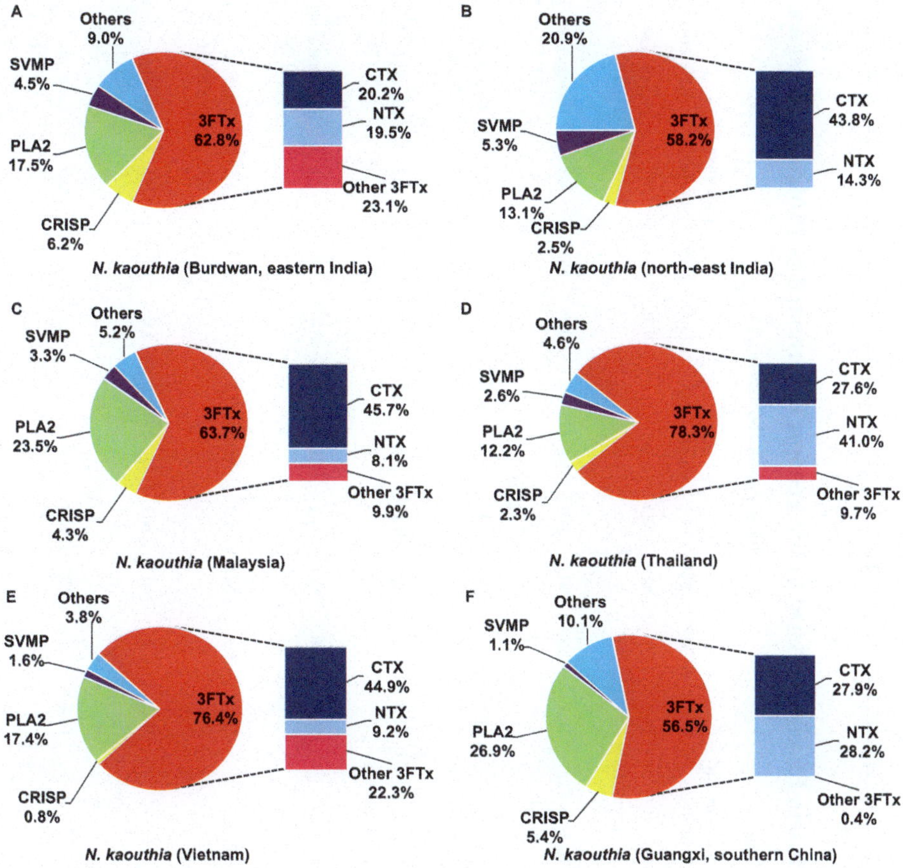

FIGURE 5.9 Venom proteome composition of *N. kaouthia* from (A) Burdwan, eastern India, (B) north-east India, (C) Malaysia, (D) Thailand, (E) Vietnam, and (F) Guangxi, southern China.

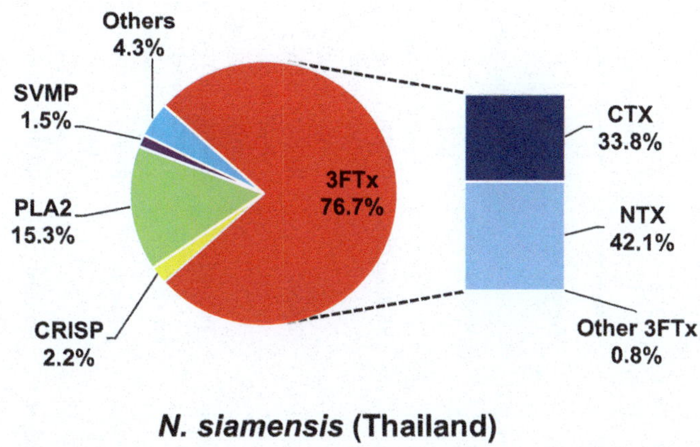

FIGURE 5.10 Venom proteome composition of *N. siamensis* from Thailand.

TABLE 5.2
A Summary of Relative Abundances of Enzymatic Protein Families in Asian Cobra Venoms Determined by Proteomic Analyses

Venom	Venom source	Proteomics analysis strategy	# proteins identified	# enzymatic proteins	Relative toxin abundance (%)									Reference
					PLA_2	SVMP	SVSP	ChE	AChE	NT	LAAO	PDE	Unknown / Non-toxin	
N. atra	Eastern Taiwan	RP-HPLC, in-solution trypsin digestion	Not mentioned	Not mentioned	13.9	2.1	ND	ND	ND	0.4	0.2	ND	ND	Huang et al. (2015)
N. atra	Western Taiwan	RP-HPLC, in-solution trypsin digestion	Not mentioned	Not mentioned	16.8	2.6	ND	ND	ND	0.5	0.2	ND	ND	Huang et al. (2015)
N. atra	Zhejiang, China	RP-HPLC, SDS-PAGE, in-gel trypsin digestion	Not mentioned	Not mentioned	12.2	1.6	ND	ND	ND	ND	ND	ND	ND	Shan et al. (2016)
N. atra	Taiwan	RP-HPLC, SDS-PAGE, in-gel trypsin digestion	Not mentioned	Not mentioned	14.4	2.7	ND	ND	ND	0.8	0.2		0.2	Liu et al. (2017)
N. kaouthia	Thailand	RP-HPLC, SDS-PAGE, in-gel trypsin digestion	38	12	13.5	2.4	ND	ND	ND	0.4	0.4	0.4	1.1	Laustsen et al. (2015)
N. kaouthia	Malaysia	RP-HPLC, SDS-PAGE, in-gel trypsin digestion	61	11	23.5	3.3	ND	ND	ND	0.3	1.1	0.4	1.6	Tan et al. (2015)
N. kaouthia	Thailand	RP-HPLC, SDS-PAGE, in-gel trypsin digestion	51	8	12.2	2.6	ND	ND	ND	0.2	1.0	0.3	0.5	Tan et al. (2015)
N. kaouthia	Vietnam	RP-HPLC, SDS-PAGE, in-gel trypsin digestion	68	8	17.4	1.6	ND	ND	ND	0.2	0.5	0.4	0.5	Tan et al. (2015)

N. kaouthia	Thailand	RP-HPLC, SDS-PAGE, in-gel trypsin digestion	Not mentioned	Not mentioned	13.5	4.4	ND	ND	ND	0.4	1.1	0.3	11.9	Liu et al. (2017)
N. kaouthia	Guangxi, southern China	RP-HPLC, SDS-PAGE, in-gel trypsin digestion	Not mentioned	Not mentioned	26.9	1.1	ND	ND	ND	ND	ND	ND	ND	Xu et al. (2017)
N. kaouthia	Burdwan, eastern India	1D-SDS-PAGE, in-gel trypsin digestion	55	21	17.5	4.5	ND	ND	0.7	0.8	1.8	0.8	0.2	Chanda et al. (2018)
N. kaouthia	Not disclosed	Whole venom in-solution trypsin digestion	Not mentioned	Not mentioned	2.3	3.4	0.1	ND	ND	0.4	0.8	0.4	0.7	Ryabinin et al. (2019)
N. kaouthia	Assam, North-east India	1D-SDS-PAGE, in-gel trypsin digestion	32	17	13.1	5.3	ND	ND	0.04	0.9	0.7	0.1	0.1	Kakati et al. (2022)
N. kaouthia	Assam, North-east India	RP-HPLC, in-solution trypsin digestion	32	12	16.0	1.5	ND	ND	0.4	ND	0.1	ND	3.8	Kakati et al. (2022)
N. naja	Rajasthan and Gujarat, north-western India	RP-HPLC, SDS-PAGE, in-gel trypsin digestion	26	2	21.4	0.9	ND	ND	ND	ND	ND	ND	0.1	Sintiprungrat et al. (2016)
N. naja	Sri Lanka	RP-HPLC, SDS-PAGE, in-gel trypsin digestion	25	2	13.9	0.8	ND	ND	ND	ND	ND	ND	0.2	Sintiprungrat et al. (2016)
N. naja	Nadia, eastern India	Cation-exchange chromatography, in-solution trypsin digestion	43	13	11.4	1.0	0.3	6.0	6.3	0.4	0.8	2.1	ND	Dutta et al. (2017)

(*Continued*)

TABLE 5.2 (CONTINUED)
A Summary of Relative Abundances of Enzymatic Protein Families in Asian Cobra Venoms Determined by Proteomic Analyses

					Relative toxin abundance (%)									
Venom	Venom source	Proteomics analysis strategy	# proteins identified	# enzymatic proteins	PLA_2	SVMP	SVSP	ChE	AChE	NT	LAAO	PDE	Unknown / Non-toxin	Reference
N. naja	Sindh, Pakistan	RP-HPLC, SDS-PAGE, in-gel trypsin digestion	55	11	14.2	1.4	ND	ND	ND	0.1	1.7	0.4	0.6	Wong et al. (2018)
N. naja	Burdwan, eastern India	1D-SDS-PAGE, in-gel trypsin digestion	52	21	20.2	6.1	ND	0.2	0.2	1.2	1.9	0.6	0.1	Chanda et al. (2018)
N. naja	Maharashtra, western India	GFC, 1D-SDS-PAGE, in-gel trypsin digestion	54	18	7.0	6.6	0.7	ND	0.06	1.4	1.2	0.8	ND	Chanda et al. (2019)
N. naja	Maharashtra, western India	RP-HPLC, SDS-PAGE, in-gel trypsin digestion	40	10	27.0	16.9	0.03	ND	ND	ND	0.3	ND	0.1	Chanda et al. (2019)
N. naja	Irula, southern India	1D-SDS-PAGE, in-gel trypsin digestion	45	21	18.2	16.2	0.01	ND	0.3	0.2	11.9	ND	ND	Chanda and Mukherjee. (2020b)
N. naja	Karnataka, southern India	RP-HPLC, in-solution trypsin digestion	101	56	6.0	8.0	0.4	ND	0.05	0.01	1.3	0.01	1.0	Attarde et al. (2021)
N. naja	Colombo, Sri Lanka	RP-HPLC, SDS-PAGE, in-gel trypsin digestion	45	14	14.6	4.0	0.9	ND	ND	0.04	0.5	0.1	1.8	Wong et al. (2021)
N. philippinensis	Southern Philippines	RP-HPLC, in-solution trypsin digestion	66	19	22.9	3.9	0.3	ND	ND	0.5	0.4	0.4	0.9	Tan et al. (2019a)

N. sagittifera	Port Blair, Andaman Island	RP-HPLC, in-solution trypsin digestion	103	58	2.1	5.0	1.1	ND	0.01	0.06	0.5	0.01	0.4	Attarde et al. (2021)
N. samarensis	Southern Philippines	RP-HPLC, in-solution trypsin digestion	31	10	3.8	4.2	ND	ND	ND	ND	0.2	ND	ND	Palasuberniam et al. (2021)
N. siamensis	Thailand	RP-HPLC, SDS-PAGE, in-gel trypsin digestion	Not mentioned	Not mentioned	15.3	1.5	ND	ND	ND	1.3	0.5	0.6	0.4	Liu et al. (2017)
N. sputatrix	Java Island	RP-HPLC, in-solution trypsin digestion	97	30	31.2	1.3	0.4	ND	0.04	0.1	0.1	0.1	0.1	Tan et al. (2017b)
N. sumatrana	Negeri Sembilan, Malaysia	RP-HPLC, in-solution trypsin digestion	32	13	34.5	1.5	0.1	ND	ND	ND	0.1	0.8	ND	Tan et al. (2022)
N. sumatrana	Penang, Malaysia	RP-HPLC, in-solution trypsin digestion	42	17	23.6	2.1	0.4	ND	0.3	0.1	0.4	0.9	0.1	Tan et al. (2022)
N. sumatrana	Southern Thailand	RP-HPLC, in-solution trypsin digestion	22	10	23.3	4.8	ND	ND	ND	ND	0.2	0.6	ND	Tan et al. (2022)
N. sumatrana	Sumatra	RP-HPLC, in-solution trypsin digestion	55	23	25.2	3.8	0.7	ND	0.1	0.1	0.7	0.5	0.2	Tan et al. (2022)

ND: not detected by mass spectrometry.

The below section describes the most common enzymes found in Asian Cobra venom samples and the pathophysiology induced by them.

5.7.1.1 Phospholipase A_2 (PLA_2)

PLA_2s (EC: 3.1.1.4) are one of the most abundant and ubiquitously expressed low molecular mass (13 to 15 kDa) enzymatic toxins (protein) in snake venoms, including those from the Asian Cobra venoms (Burke and Dennis, 2009; Dutta et al., 2015; Mukherjee, 2014a; Mukherjee et al., 2014b; Saikia et al., 2012). These enzymes catalyze the hydrolysis of membrane phospholipid molecules that contain unsaturated fatty acids at the sn-2 acyl bond, generating unsaturated fatty acids and lysophospholipids (Burke and Dennis, 2009). Notably, these hydrolysis products alter the cell membrane characteristics that promote the activation of diverse signaling pathways, ultimately culminating in extensive cellular pathology (Figure 5.11). Pharmacological characterization of several Cobra venom PLA_2s has closely linked them to anticoagulant action, necrosis, myotoxicity, cardiotoxicity, neurotoxicity, hemorrhage, hemolysis, edema, and hypotension (Dennis et al., 2011; Doley et al., 2004; Doley and Mukherjee, 2003; Dutta et al., 2015; Mukherjee et al., 2014b). Moreover, these enzymes act synergistically with other Cobra venom components, including cytotoxins, to potentiate cell membrane damage and tissue necrosis (Dutta et al., 2019).

PLA_2s constituted 2.2 to 34.5% of the total proteome of Asian Cobra venoms (Table 5.2, Figure 5.12) with several isoforms ranging from 1 to 9; however, astonishingly enough, no PLA_2 was identified in *N. oxiana* from northern India perhaps

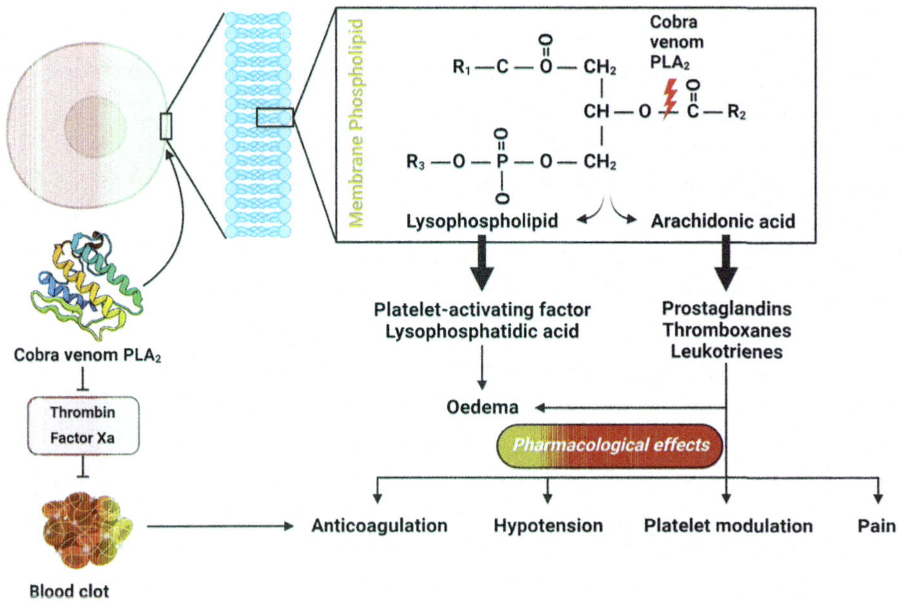

FIGURE 5.11 A schematic diagram to represent the catalysis mechanisms of Cobra venom PLA_2 enzymes and the subsequent pharmacological complications.

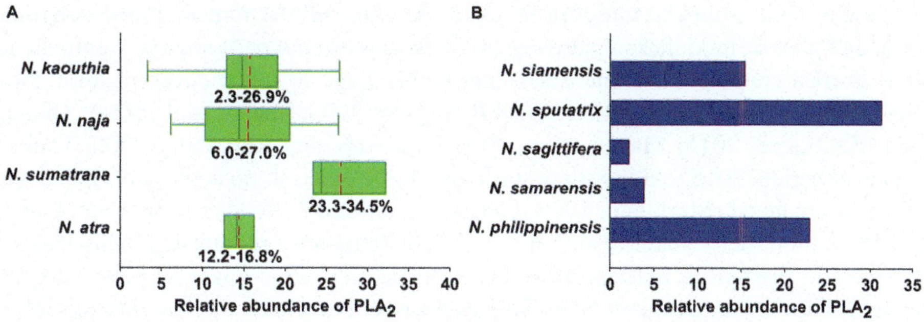

FIGURE 5.12 Relative distribution of PLA_2 in the venoms of **A.** *N. kaouthia, N. sumatrana, N. naja*, and *N. atra* from multiple geographic locations, and **B.** *N. philippinensis, N. sagittifera, N. samarensis, N. siamensis*, and *N. sputatrix* from a single geographic location.

due to a lack of venom decomplexation strategy in the proteomic workflow (Deka et al., 2023). Interestingly, the relative abundance of PLA_2 was exceptionally low in the venoms of *N. sagittifera* (2.1%) from Andaman Island, *N. samarensis* (3.8%) from southern Philippines, and *N. kaouthia* (2.3%) of unknown geographic origin.

In addition, the abundance of this ubiquitous class of toxin was relatively lower in *N. naja* samples from Maharashtra, western India (7.0%) and Karnataka, southern India (6.0%). Conversely, among the remaining Asian Cobra venom proteomes profiled, *N. sumatrana* from Negeri Sembilan, Malaysia (34.5%) and *N. sputatrix* from Java Island (31.2%) were reported with the highest relative abundance of PLA_2s (Table 5.2). It is noteworthy to mention that a recent study demonstrated that venoms of Asian spitting Cobras (*N. sumatrana* and *N. sputatrix*) exhibit significantly higher PLA_2 activity as compared to the non-spitting species (*N. atra*, *N. naja*, and *N. kaouthia*) (Tan et al., 2019b). Furthermore, the increased expression of PLA_2s in spitting Cobra species to potentiate cytotoxin activity has been postulated as convergent evolution of pain-inducing defensive venoms (Kazandjian et al., 2021). Therefore, these supporting arguments indicate that these discrepancies in the relative abundances of PLA_2s might be due to evolution-driven inter-specific and intra-specific geographic variations in Asian Cobra venoms. Nevertheless, the contribution of different analytical workflows, mass spectrometers, and search engines on minor variations in the relative abundance of toxins cannot be ruled out.

5.7.1.2 Snake Venom Metalloprotease (SVMP)

SVMPs (20 to 110 kDa; EC: 3.4.24.-) are a sub-class of proteolytic enzymes that can cleave specific proteins related to blood coagulation cascade, platelet aggregation pathways, and extracellular matrix (Fox and Serrano, 2008; Gutierrez and Rucavado, 2000; Mukherjee, 2008; Oliveira et al., 2010; Thakur et al., 2015). They are usually multi-domain proteins requiring divalent metal ions such as Mg^{2+} and Zn^{2+} to catalyze their activity. Structurally, they are classified under P-I, P-II, and P-III classes; P-I SVPMs contain the catalytic domain only, P-II SVMPs contain a disintegrin domain in addition to the catalytic domain, and P-III SVMPs comprise catalytic, disintegrin and cysteine-rich domains (Fox and Serrano, 2008; Matsui et al., 2000).

Due to their ability to target proteins of the extracellular matrix, blood coagulation, and platelet modulation pathways, they have profound pathological significance in disturbed coagulation homeostasis and hemorrhage in snake envenomation, particularly in viperid bites (Gutierrez and Rucavado, 2000; Matsui et al., 2000; Thakur and Mukherjee, 2015; Matsui et al., 2000; Gutierrez and Rucavado, 2000; Thakur and Mukherjee, 2015). Along these lines, the fibrin(ogen)olytic activity of SVMPs is more pronounced (Markland, 1998; Sanchez et al., 2017). However, in Asian Cobra envenomation, hemostatic disturbances are not frequently manifested. This lack of hemostatic disruption activity in Asian Cobras, like other elapids, is due to the expression of lower amounts of SVMPs in their venom glands than in viperid snakes (Table 5.2). Nevertheless, a few SVMPs with fibrin(ogen)olytic and platelet inhibiting activities have been purified and characterized from different Asian Cobra species (Hamako et al., 1998; Jagadeesha et al., 2002; Kumar et al., 2010; Sun and Bao, 2010; Sun et al., 2020; Wijeyewickrema et al., 2007).

A variable number of SVMP isoforms (1 to 19) were detected in all the Asian Cobra venom proteomes. Their relative abundances were determined in the range of 0.8 to 16.9% (Table 5.2, Figures 5.1–5.10, 5.13), the highest being found in *N. naja* venom samples from western India (16.9%) when profiled by the venomics approach and southern India (16.2%) that was characterized by the gel-based approach (Chanda et al., 2019; Chanda and Mukherjee, 2020b). Interestingly, when *N. naja* venom proteome from western India was deciphered using a complementary approach involving GFC of the crude venom followed by SDS-PAGE analysis of the venom fractions and subsequent in-gel trypsin digestion and mass spectrometry analysis, the relative abundance of SVMPs was found to be 6.6%. This finding highlights that different analytical methods may yield varied relative quantities for the identified proteins in the same venom sample. Subsequently, as a form of validation, the band intensities of the crude venom were analyzed and correlated with the relative abundance of the toxin classes identified by both proteomic approaches (GFC followed by SDS-PAGE, in-gel trypsin digestion and MS/MS analysis, and venomics approach). While the complementary approach was more consistent with the SDS-PAGE band intensity of crude *N. naja* venom, the relative abundances calculated by the venomics approach seemed to have deviated significantly from band intensity values (Chanda et al., 2019).

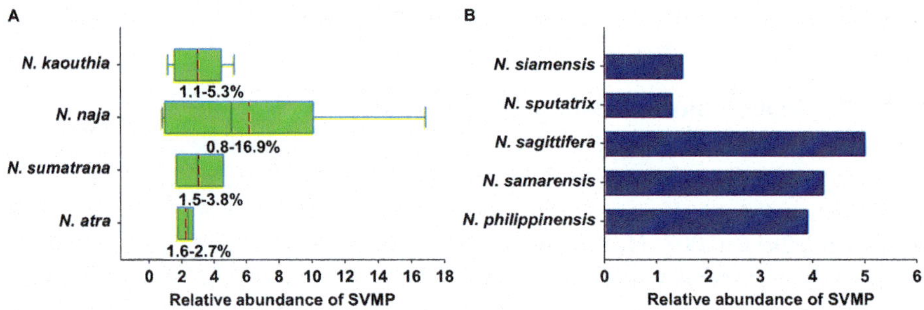

FIGURE 5.13 Relative distribution of SVMP in the venoms of **A.** *N. kaouthia, N. sumatrana, N. naja*, and *N. atra* from multiple geographic locations, and **B.** *N. philippinensis, N. sagittifera, N. samarensis, N. siamensis*, and *N. sputatrix* from a single geographic location.

Nevertheless, the relative abundances of SVMPs in the same Asian Cobra species (*N. naja*) from Sri Lanka and north-western India were determined at only 0.8% and 0.9%, respectively, indicating significant intra-specific venom variation due to geographical location. Apart from *N. naja* venom proteome, the variation in the relative abundance of SVMPs among other Asian Cobra species was found to be marginal; *N. kaouthia* (1.1 to 5.3%), *N. siamensis* (1.5%), *N. sumatrana* (1.5 to 4.8%), *N. sputatrix* (1.3%), *N. sagittifera* (5.0%), *N. samarensis* (4.2%), *N. philippinesis* (3.9%) and *N. atra* (1.6 to 2.7%) (Table 5.2).

5.7.1.3 Snake Venom Serine Protease (SVSP)

SVSPs (~25 to 65 kDa; EC: 3.4.21.-) belonging to the S1 family of serine proteases are multifunctional enzymatic proteins that can specifically cleave polypeptide chains at the C-terminal end of hydrophobic or positively charged amino acid residues (Mackessy, 2010; Mukherjee, 2014b; Mukherjee and Mackessy, 2013; Siigur et al., 2003). Like SVMPs, they are abundant and, therefore, widely characterized in viperid venoms that exhibit prominent hemostatic disturbances in envenomed victims. Usually, they have a pronounced effect on the hemostatic system. They can exert either pro-coagulant effects including factor V activation, platelet aggregation, fibrin formation, thrombin-like activity, kininogenolysis, or anticoagulant effects including protein C activation, fibrinolysis and plasminogen activation (Kini, 2006; Mackessy, 2010; Matsui et al., 2000; Mukherjee, 2014b; Mukherjee and Mackessy, 2013; Siigur et al., 2003).

The Cobra venom contains a deficient proportion of SVSP. Variable number of SVSPs (1 to 15) were identified in venoms of *N. philippinensis, N. sagittifera, N. sputatrix*, and a few samples of *N. sumatrana* and *N. naja* with relative abundances ranging from 0.01 to 1.1% only (Table 5.2). Consequently, they can be considered minor enzymatic toxins in Asian Cobra venoms.

5.7.1.4 Nucleotidase (NT)

Nucleotidases (ATPase, ADPase, apyrase, and 5'-nucleotidase) are a class of monomeric high molecular weight (~50 to 80 kDa) hydrolytic enzymes of snake venoms that act on nucleic acids and/or their derivatives to generate purines and pyrimidines (Dhananjaya and D'Souza, 2010; Mackessy, 2002; Ogawa et al., 2009; Trummal et al., 2015). ATPases (EC: 3.6.1.3) hydrolyse ATP, ADPases (EC: 3.6.1.5) hydrolyse ADP, apyrases (EC: 3.6.1.5) hydrolyse both ATP and ADP, and 5'-NTs (EC: 3.1.3.5) hydrolyse ribose and deoxyribose monophosphates (Figure 5.14) (Dhananjaya and D'Souza, 2010; Kalita et al., 2018c).

Although nucleotidases are ubiquitously expressed in venom glands, their pharmacological role remains poorly understood. Nevertheless, several studies have shown that the depletion of ATP by its hydrolysis may be one of the mechanisms that potentiate prey immobilization by the nucleotidases. Moreover, the purines and pyrimidine bases generated by the nucleotidases may also have several pharmacological effects, such as vasodilation, edema, inhibition of platelet aggregation, and renal complications (Kalita et al., 2018c). Notably, 5'-NTs remain the most studied class among nucleotidases, while the primary sequence of the remaining enzyme sub-classes is yet to be determined. Therefore, it is logical that

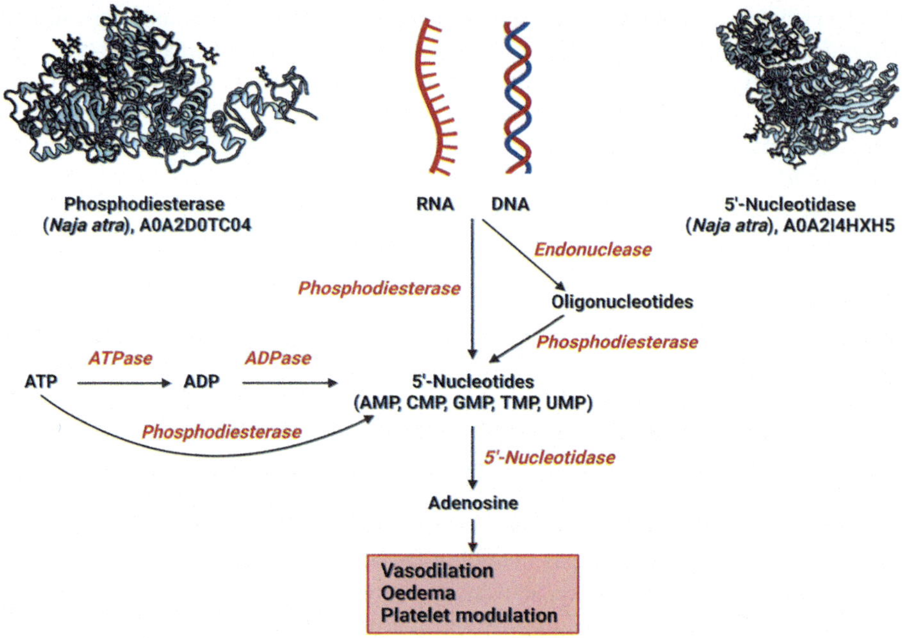

FIGURE 5.14 Catalysis of nucleic acids, nucleotides, ATP, and ADP by Cobra venom PDEs, 5′-NTs, ATPases, and ADPases.

database-dependent proteomic analysis could only identify 5'-NTs in the venoms of Asian Cobras even though the presence of other sub-classes has been demonstrated by biochemical studies (Chanda et al., 2019; Chanda and Mukherjee, 2020b; Chanda et al., 2018; Dutta et al., 2017).

The nucleotidase activity of elapid venoms is relatively lower compared to viperid and crotalid venoms. Consequently, they are found in minor amounts in Asian Cobra venoms (1 to 3 isoforms with relative abundance ranging from 0.01 to 1.4%) (Table 5.2). However, 5′-NTs could not be identified in venom samples of *N. atra* from Zhejiang, China, *N. samarensis* from southern Philippines, *N. sumatrana* from Malaysia and south Thailand, *N. naja* from western India, north-western India and Sri Lanka, and *N. kaouthia* from south China, perhaps due to their significantly lower abundance in these venoms which is beyond the detection limit of the analytical workflow.

5.7.1.5 Phosphodiesterase (PDE)

PDE (EC: 3.1.4.1) is yet another monomeric high molecular weight (~90 to 100 kDa) hydrolytic enzyme specifically targeting the phosphodiester bonds present in polynucleotides. The catalysis of polynucleotides by PDEs is sequential; they start cleaving from the 3' end and generate 5′-mononucleotides. In addition, similar to nucleotidases, they can also target nucleic acid derivatives, including ATP, GTP, CTP, etc. (Figure 5.14) (Fox, 2013; Santoro et al., 2009; Valerio et al., 2002).

PDEs are also distributed ubiquitously in snake venoms but are yet to be thoroughly characterized. A few studies suggest that PDEs can hydrolyze DNA and RNA molecules, thereby releasing purines that have subsequent pathological outcomes in prey/victims similar to nucleotidases (Santoro et al., 2009; Valerio et al., 2002). PDEs are considered minor components of snake venoms, and viperid venoms usually contain more PDEs than elapid venoms (Oliveira et al., 2022). PDEs were identified in almost all Asian Cobra species (1 to 3 isoforms) except in the venoms of *N. atra. N. samarensis* and a few samples of *N. naja* and *N. kaouthia*. Furthermore, their relative abundance demonstrated minor variations among Asian Cobra species (0.01 to 2.1%) (Table 5.2).

5.7.1.6 L-Amino Acid Oxidase (LAAO)

Snake venom LAAOs (EC: 1.4.3.2) are a class of FAD-dependent oxidoreductase enzymes that can specifically catalyze the oxidative deamination of L-amino acids to α-keto acids. They are usually high molecular weight snake venom enzymes and can be either monomeric (50 to 70 kDa) or homodimeric (120 to 150 kDa) (Du and Clemetson, 2002; Mukherjee et al., 2015). Notably, snake venoms' pale yellow color is primarily due to the flavin group (FAD) in LAAOs (Mukherjee et al., 2015).

The catalytic mechanism of these oxidoreductases includes two half-reactions. The first half is a reduction reaction that converts FAD into $FADH_2$ coupled with concomitant oxidation of L-amino acids into imino acids, which are ultimately hydrolyzed into ammonia and α-keto acids. The second half is an oxidation reaction involving the conversion of $FADH_2$ into FAD in the presence of oxygen, leading to hydrogen peroxide generation (Figure 5.15) (Naumann et al., 2011). Pharmacologically, these enzymes can cause significant cytotoxicity due to the

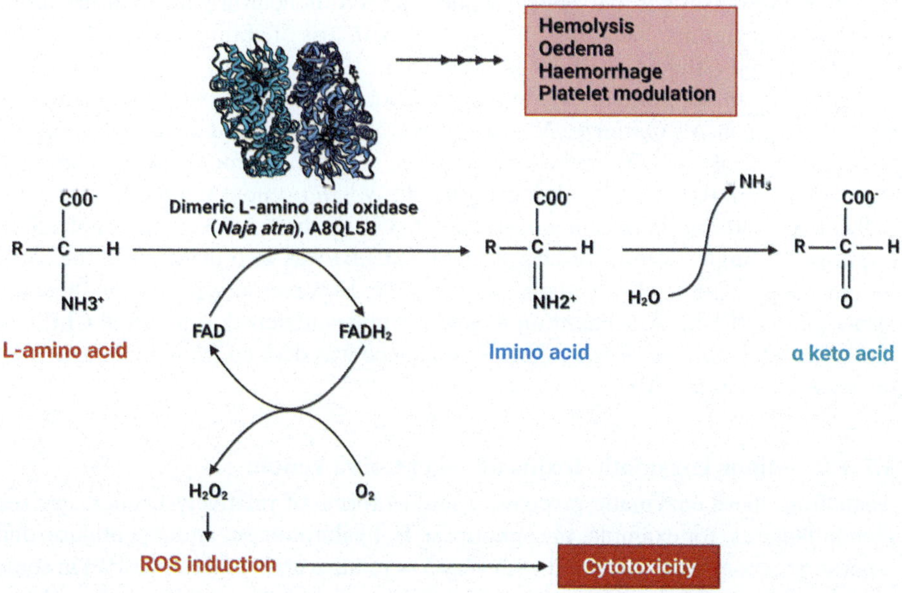

FIGURE 5.15 Catalytic mechanism of Cobra venom LAAO and its associated pharmacological effects.

production of hydrogen peroxide as well as contribute to hemolysis, edema, hemorrhage, and disruption of platelet function by diverse other mechanisms (Fox, 2013; Hiu and Yap, 2020; Mukherjee et al., 2015).

LAAO isoforms (1 to 7) were identified in all the Asian Cobra venoms except for a couple of *N. naja* venom samples and one venom sample each of *N. kaouthia* and *N. atra* (Table 5.2). The relative abundance of LAAOs marginally varied between 0.1 to 1.9% in most of the Asian Cobra venom samples; however, for one venom sample of *N. naja* from Irula, southern India, the relative abundance of LAAO was unusually high (11.8%) (Table 5.2). Although such high amounts of LAAO have not been reported in venoms of *Naja* sp., More et al. (2010) purified a 55 kDa LAAO from the venom of *Bungarus caeruleus* (an elapid), constituting ~25% of the total venom (More et al., 2010). Interestingly, it has also been observed in *Crotalus viridis* that under certain circumstances, one venom gland of an individual can produce yellow venom with more amounts of LAAO. In contrast, the other gland from the same individual produces a colorless toxin with undetectable LAAO (Smith and Mackessy, 2020).

5.7.1.7 Acetylcholinesterase (AChE)

AChE (~60 to 70 kDa) is a serine hydrolase enzyme class that can catalyze acetylcholine (a neurotransmitter) breakdown into choline and acetate. AChEs (EC: 3.1.1.7) have been purified primarily from elapid venoms (Ahmed et al., 2009; Frobert et al., 1997); however, they have not been thoroughly characterized in the context of pathophysiological effects of envenomation. Nevertheless, owing to their catalytic property, it has been suggested that they are responsible for prompt hydrolysis of acetyl choline neurotransmitter that can subsequently result in neurological manifestations in envenomed victims, including blurring of vision, loss of ability to focus, and dilation of the pupil (Ahmed et al., 2009; Raba et al., 1979).

AChEs are minor components in Asian Cobra venoms, and one or two isoforms were detected in *N. sputatrix*, *N. sumatrana*, and a few samples of *N. naja* and *N. kaouthia* (Table 5.2). The relative abundance of AChEs in most of the Cobra venoms varied from 0.04 to 0.7%; however, for the *N. naja* venom sample from eastern India, two isoforms of AChE constituted 6.3% of the total venom (Table 5.2). Moreover, a single isoform of cholinesterase (ChE) was also detected in the same venom sample with a relative abundance of 6.0%. However, it is to be noted that the identification of ChE was based on MS/MS peptides matched to a partial ChE isoform from *Ophiophagus hannah*, whose sequence was derived from genome assembly data (Dutta et al., 2017).

5.7.1.8 Minor Enzymatic Toxins of Asian Cobra Venom

Apart from these enzymatic proteins, a few isoforms of relatively uncharacterized protein families, for example, phospholipase B, hyaluronidase, aminopeptidase, and aspartic protease, were identified with very low relative abundances (<0.1%) in some venom samples of Asian Cobras suggesting they might have certain ancillary functions. Nevertheless, the exact role of these enzymes in envenomation has not been elucidated.

5.7.2 Distribution of Non-Enzymatic Toxins in Diverse Asian Cobra Venoms

As for most elapids, non-enzymatic proteins predominate in Asian Cobra venoms. These non-enzymatic toxins have diverse mechanisms of action, and they exhibit profound pathophysiological manifestation upon Cobra envenomation. The non-enzymatic protein families identified in various Asian Cobra venoms include three-finger toxins (3FTx), cysteine-rich secretory protein (CRISP), Cobra venom factor (CVF), nerve growth factor (NGF), vespryn (VES) or Ohanin-like protein (OLP), cystatin (CYS), Kunitz-type serine protease inhibitor (KSPI), natriuretic peptide (NP), vascular endothelial growth factor (VEGF), and C-type lectin (CTL). 3FTx was the most abundant non-enzymatic protein family in all the Asian Cobra venoms, followed by CRISP. An account of the non-enzymatic proteins and relative abundances of toxin classes identified in the Asian Cobra venom proteomes is summarized in Table 5.3.

5.7.2.1 Three-Finger Toxin (3FTx)

3FTxs are low molecular weight (~6 to 9 kDa) non-enzymatic proteins predominantly found in elapid and colubrid venoms. They contain 60–74 amino acid residues and four conserved disulfide linkages (Dufton and Hider, 1983; Utkin, 2013, 2019). Structurally, they have a small hydrophobic core stabilized by the four conserved disulfide linkages and three β-stranded loops or 'fingers' protruding from the core. Based on their pharmacological actions, Cobra venom 3FTxs can be broadly classified into alpha-neurotoxins (αNTX) and cytotoxins (CTX). Further, based on their length and number of disulfide bridges, the αNTX sub-class is further divided into long-chain αNTX (65 to 72 amino acids with five disulfide bridges) and short-chain αNTX (60 to 62 amino acids with four disulfide bridges) (Kini and Doley, 2010; Utkin, 2013). While both types of αNTX can specifically bind to the muscle nicotinic acetylcholine receptor (nAChR) and cause neuromuscular paralysis, the effect of CTXs is largely non-specific and causes local tissue necrosis.

In addition, a few CTXs have also been reported to cause depolarization of heart muscle membranes that results in cardiac failures, and thus, they are known as cardiotoxins (CdTx) (Slotta and Vick, 1969). Apart from these two major categories, other types of 3FTxs include muscarinic toxin-like protein (MTLP) and weak toxins (WTX) that usually occur in minor amounts in Cobra venoms (Utkin, 2013). It is to be noted that Cobra envenomation primarily manifests neurotoxicity and local tissue damage of variable degrees, and different sub-classes of 3FTxs are responsible for these pathophysiological complications. Therefore, 3FTxs represent one of the major toxin classes in Asian Cobra venoms.

With 10 to 24 isoforms identified by proteomic analyses, 3FTxs constituted 37.4 to 85.7% of the total proteome of Asian Cobra venoms; the venom proteomes of *N. naja* (southern India) and *N. kaouthia* of an unknown geographic location were reported with the lowest and highest proportions of 3FTxs, respectively (Table 5.3, Figures 5.2–5.10, 5.16). More importantly, there was significant variation in the relative proportion of CTXs and αNTXs in the venom proteomes (Figures 5.17 and 5.18). The venoms of *N. atra*, *N. sumatrana*, *N. sputatrix*, *N. naja* from Sri Lanka,

TABLE 5.3
A Summary of Relative Abundances of Non-Enzymatic Protein Families in Asian Cobra Venoms Determined by Proteomic Analyses

					Relative Toxin Abundance (%)												
Venom	Venom Source	Proteomics Analysis Strategy	# Proteins Identified	# Non-enzymatic Proteins	CTX	NTX	Other 3FTx	CRISP	NGF	CVF	Ves / OLP	KSPI	CYS	NP	VEGF	CTL	Reference
N. atra	Eastern Taiwan	RP-HPLC, in-solution trypsin digestion	Not mentioned	Not mentioned	59.4	20.5	ND	2.2	1.1	0.2	ND	ND	ND	ND	ND	ND	Huang et al. (2015)
N. atra	Western Taiwan	RP-HPLC, in-solution trypsin digestion	Not mentioned	Not mentioned	52.9	23.5	ND	2.4	0.9	0.2	ND	ND	ND	ND	ND	ND	Huang et al. (2015)
N. atra	Zhejiang, China	RP-HPLC, SDS-PAGE, in-gel trypsin digestion	Not mentioned	Not mentioned	65.3	11.2	7.8	1.8	0.1	ND	ND	ND	ND	ND	ND	ND	Shan et al. (2016)
N. atra	Taiwan	RP-HPLC, SDS-PAGE, in-gel trypsin digestion	Not mentioned	Not mentioned	53.0	15.1	0.9	6.0	2.9	ND	3.8	ND	ND	ND	ND	ND	Liu et al. (2017)
N. kaouthia	Thailand	RP-HPLC, SDS-PAGE, in-gel trypsin digestion	38	26	24.3	53.2	ND	1.7	0.3	0.1	0.6	ND	ND	1.1	ND	0.5	Laustsen et al. (2015)
N. kaouthia	Malaysia	RP-HPLC, SDS-PAGE, in-gel trypsin digestion	61	50	45.7	8.1	9.9	4.3	0.2	0.8	0.3	0.5	ND	ND	ND	ND	Tan et al. (2015)
N. kaouthia	Thailand	RP-HPLC, SDS-PAGE, in-gel trypsin digestion	51	43	27.6	41.0	9.7	2.3	0.5	1.1	0.7	0.1	ND	0.2	ND	ND	Tan et al. (2015)
N. kaouthia	Vietnam	RP-HPLC, SDS-PAGE, in-gel trypsin digestion	68	60	44.9	9.2	22.3	0.8	1.3	0.7	0.2	ND	ND	ND	ND	ND	Tan et al. (2015)

N. kaouthia	Thailand	RP-HPLC, SDS-PAGE, in-gel trypsin digestion	Not mentioned	Not mentioned	20.9	40.8	0.4	2.8	0.8	0.4	2.3	ND	ND	ND	ND	ND	Liu et al. (2017)
N. kaouthia	Guangxi, southern China	RP-HPLC, SDS-PAGE, in-gel trypsin digestion	Not mentioned		27.9	28.2	0.4	5.4	1.0	ND	9.2	ND	ND	ND	ND	ND	Xu et al. (2017)
N. kaouthia	Burdwan, eastern India	1D-SDS-PAGE, in-gel trypsin digestion	55	34	20.2	19.5	23.1	6.2	1.2	2.3	1.2	ND	ND	ND	ND	ND	Chanda et al. (2018)
N. kaouthia	Not disclosed	Whole venom in-solution trypsin digestion	Not mentioned	Not mentioned		85.7		1.2	0.9	1.9	2.0	ND	0.01	0.1	ND	0.1	Ryabinin et al. (2019)
N. kaouthia	Assam, North-east India	1D-SDS-PAGE, in-gel trypsin digestion	32	15	43.8	14.3	ND	2.5	5.9	0.6	3.6	ND	8.8	ND	ND	ND	Kakati et al. (2022)
N. kaouthia	Assam, North-east India	RP-HPLC, in-solution trypsin digestion	32	20	37.4	12.3	14.4	0.1	2.1	0.9	2.5	ND	ND	8.1	ND	ND	Kakati et al. (2022)
N. naja	Sri Lanka	RP-HPLC, SDS-PAGE, in-gel trypsin digestion	25	23	71.5	6.2	0.2	3.7	0.6	ND	ND	0.3	ND	ND	ND	ND	Sintiprungrat et al. (2016)
N. naja	Rajasthan and Gujarat, north-western India	RP-HPLC, SDS-PAGE, in-gel trypsin digestion	25	24	69.3	3.7	1.1	2.5	0.7	ND	ND	0.1	ND	ND	ND	ND	Sintiprungrat et al. (2016)
N. naja	Nadia, eastern India	Cation-exchange chromatography, in-solution trypsin digestion	43	30		63.8		2.1	0.9	1.1	1.3	0.4	ND	2.0	ND	ND	Dutta et al. (2017)

(Continued)

TABLE 5.3 (CONTINUED)

A Summary of Relative Abundances of Non-Enzymatic Protein Families in Asian Cobra Venoms Determined by Proteomic Analyses

Venom	Venom Source	Proteomics Analysis Strategy	# Proteins Identified	# Non-enzymatic Proteins	Relative Toxin Abundance (%)												Reference
					CTX	NTX	Other 3FTx	CRISP	NGF	CVF	Ves / OLP	KSPI	CYS	NP	VEGF	CTL	
N. naja	Sindh, Pakistan	RP-HPLC, SDS-PAGE, in-gel trypsin digestion	55	44	46.9	26.3	2.1	0.7	0.7	0.6	3.4	0.9	ND	ND	ND	ND	Wong et al. (2018)
N. naja	Burdwan, eastern India	1D-SDS-PAGE, in-gel trypsin digestion	52	31	24.1	21.6	15.4	3.0	3.1	1.7	0.5	ND	0.1	ND	ND	ND	Chanda et al. (2018)
N. naja	Maharashtra, western India	RP-HPLC, SDS-PAGE, in-gel trypsin digestion	40	30		50.9		1.1	1.3	0.03	1.3	1.0	0.01	ND	ND	0.1	Chanda et al. (2019)
N. naja	Maharashtra, western India	GFC, 1D-SDS-PAGE, in-gel trypsin digestion	54	36		68.5		2.9	1.1	0.6	3.8	4.0	0.4	ND	ND	ND	Chanda et al. (2019)
N. naja	Irula, southern India	1D-SDS-PAGE, in-gel trypsin digestion	45	24		37.4		4.7	0.6	4.3	1.6	1.4	0.3	2.9	ND	ND	Chanda and Mukherjee. (2020b)
N. naja	Karnataka, southern India	RP-HPLC, in-solution trypsin digestion	101	45	10.1	51.4	ND	11.5	0.3	0.1	6.3	1.4	2.1	ND	ND	ND	Attarde et al. (2021)
N. naja	Colombo, Sri Lanka	RP-HPLC, SDS-PAGE, in-gel trypsin digestion	45	31	56.9	10.5	4.8	3.1	0.7	1.7	0.3	0.1	ND	ND	ND	ND	Wong et al. (2021)
N. philippinensis	Southern Philippines	RP-HPLC, in-solution trypsin digestion	66	47	21.3	44.5	0.8	1.5	0.1	2.4	0.1	ND	ND	ND	ND	ND	Tan et al. (2019a)

N. sagittifera	Port Blair, Andaman Island	RP-HPLC, in-solution trypsin digestion	103	45	11.6	57.9	ND	8.3	0.1	2.6	8.2	0.05	1.8	0.1	0.1	0.04	Attarde et al. (2021)
N. samarensis	Southern Philippines	RP-HPLC, in-solution trypsin digestion	31	21	16.3	65.9	8.3	1.1	0.1	ND	0.1	ND	ND	ND	ND	ND	Palasuberniam et al. (2021)
N. siamensis	Thailand	RP-HPLC, SDS-PAGE, in-gel trypsin digestion	Not mentioned	Not mentioned	33.8	42.1	0.8	2.2	0.7	0.3	0.5	ND	ND	ND	ND	ND	Liu et al. (2017)
N. sputatrix	Java Island	RP-HPLC, in-solution trypsin digestion	97	67	48.0	10.7	5.5	ND	1.8	0.3	0.1	0.2	ND	ND	ND	ND	Tan et al. (2017b)
N. sumatrana	Negeri Sembilan, Malaysia	RP-HPLC, in-solution trypsin digestion	32	19	30.0	14.8	13.7	0.4	2.8	1.2	ND	ND	ND	ND	ND	ND	Tan et al. (2022)
N. sumatrana	Penang, Malaysia	RP-HPLC, in-solution trypsin digestion	42	25	52.8	5.4	5.6	0.2	6.6	1.4	0.1	ND	ND	ND	ND	ND	Tan et al. (2022)
N. sumatrana	Southern Thailand	RP-HPLC, in-solution trypsin digestion	22	12	53.8	16.0	ND	ND	0.9	0.3	0.1	ND	ND	ND	ND	ND	Tan et al. (2022)
N. sumatrana	Sumatra	RP-HPLC, in-solution trypsin digestion	55	32	33.1	10.8	13.8	0.5	3.2	1.4	0.1	5.8	ND	ND	ND	ND	Tan et al. (2022)

ND: not detected by mass spectrometry.

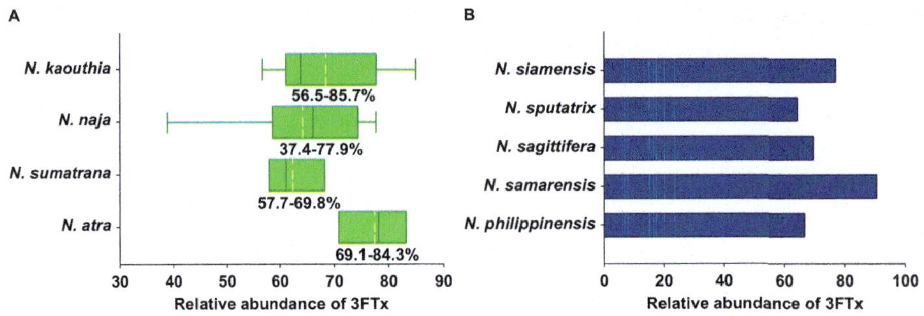

FIGURE 5.16 Relative distribution of 3FTx in the venoms of (A) *N. kaouthia, N. sumatrana, N. naja*, and *N. atra* from multiple geographic locations, and (B) *N. philippinensis, N. sagittifera, N. samarensis, N. siamensis*, and *N. sputatrix* from a single geographic location.

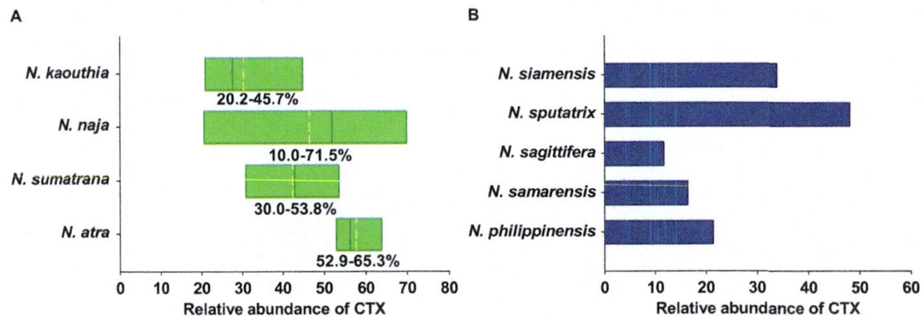

FIGURE 5.17 Relative distribution of CTX in the venoms of (A) *N. kaouthia, N. sumatrana, N. naja*, and *N. atra* from multiple geographic locations, and (B) *N. philippinensis, N. sagittifera, N. samarensis, N. siamensis*, and *N. sputatrix* from a single geographic location.

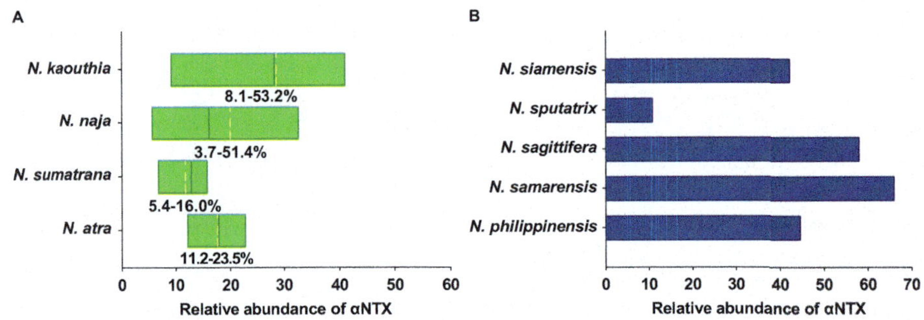

FIGURE 5.18 Relative distribution of αNTX in the venoms of (A) *N. kaouthia, N. sumatrana, N. naja*, and *N. atra* from multiple geographic locations, and (B) *N. philippinensis, N. sagittifera, N. samarensis, N. siamensis*, and *N. sputatrix* from a single geographic location.

Pakistan, and north-western India, and *N. kaouthia* from Malaysia and Vietnam contained ~2 to 18 folds more CTXs compared to αNTXs. On the contrary, the relative abundance of αNTXs in the venoms of *N. philippinensis, N. samarensis, N. saggitifera, N. naja* from southern India, and *N. kaouthia* from Thailand was ~1.5 to 5 folds higher than CTXs. Furthermore, a few venom samples, including *N. siamensis, N. naja* from eastern India, and *N. kaouthia* from China and east India, contained comparable amounts of CTXs and αNTXs.

Increased expression of CTXs in Cobra venoms has been hypothesized as a major evolutionary adaptation for the spitting behavior of certain Cobra species, and consequently, spitting Cobras contain relatively higher proportions of CTXs (Kazandjian et al., 2021). Among the Asian species, *N. philippinensis, N. samarensis, N. siamensis, N. sumatrana*, and *N. sputatrix* have been reported to spit behavior. Therefore, it is logical that the relative abundance of CTXs in *N. sumatrana* and *N. sputatrix* venoms is comparatively higher than that of αNTXs. However, such a discrete correlation between spitting behavior and relative abundance of CTX was not apparent in the venoms of the rest of the spitting species (*N. philippinensis, N. samarensis*, and *N. siamensis*) as well as the non-spitting species (*N. atra, N. naja*, and *N. kaouthia*).

It is to be noted that while the relative abundance of αNTXs correlates well with the severity of neurotoxicity observed in Cobra-envenomed patients, the same is not valid for CTXs (Panagides et al., 2017; Tan et al., 2019a; Tan et al., 2017b). Asian Cobra venoms contain CTXs with variable potencies; thus, a distinct correlation between their relative abundance and severity of local tissue-damaging effects is not always observed. A similar explanation can also be implicated for the observed discrepancy between the relative abundance of CTXs and spitting behavior. Nevertheless, these prominent variations in the relative abundance of 3FTxs and its sub-classes signify inter-specific and intra-specific geographic variations in Asian Cobra venoms.

5.7.2.2 Cysteine-Rich Secretory Protein (CRISP)

CRISPs are a group of mid-molecular weight (~20 to 30 kDa) non-enzymatic proteins widespread in venoms of Elapidae, viperidae, and colubridae snake families (Yamazaki and Morita, 2004). As their name suggests, they are rich in cysteine residues; more specifically, they contain 16 conserved cysteine residues that take part in the formation of 8 disulfide linkages, out of which 4 of them towards the C-terminal end forms the cysteine-rich domain (CRD region) (Tadokoro et al., 2020). To date, the biological activity of the majority of CRISPs is restricted to ion channel blockers that can inhibit smooth muscle contraction (Brown et al., 1999; Osipov et al., 2005). Nevertheless, the discrete functional roles of CRISPs in the pathophysiology of Cobra envenomation remain unknown.

A variable number of CRISPs (1 to 7) were detected in almost all the Asian Cobra venom proteomes, with the exceptions of *N. sputatrix* (Java) and *N. sumatrana* (Thailand) (Table 5.3). While the relative abundance of CRISPs in most of the venom samples ranged between 0.2 to 6.2%, two venom samples, *N. sagittifera* and *N. naja* from Karnataka, southern India, contained relatively higher proportions of CRISPs (8.3 and 11.5%, respectively) (Table 5.3). These disparities might be because of inter-specific (in the case of *N. sagittifera*) and geographical (in the case of *N. naja*) venom variations; however, the implication of such differences in the severity of envenomation can be challenging to comprehend since the contribution of CRISPs in pathophysiology is not deciphered yet.

5.7.2.3 Nerve Growth Factor (NGF)

NGFs are a class of neurotrophins that can regulate the growth and differentiation of sensory and motor neurons. Snake venom NGFs can exist as monomers, homo- or hetero-dimers, and the molecular weight of the monomers ranges from 12 to 20 kDa (Islam et al., 2020; Trummal et al., 2011). Although they are ubiquitously present in the venoms of the Viperidae, Elapidae and Colubridae families of snakes, and their neuritogenesis capability is well established (Islam et al., 2021; Islam et al., 2020; Kostiza and Meier, 1996), their exact role in venom-induced pathophysiology is not fully understood. Nevertheless, a recent intriguing study on a cognate complex (comprising of PLA_2, 3FTxs, and NGF) isolated from eastern India Cobra *N. naja* venom suggested that NGFs can promote the binding and cytotoxicity of the cognate complex, thereby potentiating the pharmacological effects of the individual components (Dutta et al., 2019). Another interesting study from our lab demonstrated the neuritogenesis (NGF-like property) potency of a long-chain α-neurotoxin (Nn-α-elapitoxin-1) isolated from *N. naja* venom (Islam et al., 2020). Interestingly, Nn-α-elapitoxin-1 did not exhibit any sequence similarity to existing snake venom NGFs and yet could induce neuritogenesis in rat pheochromocytoma (PC-12) cells by high-affinity binding to the tyrosine kinase A (TrkA) receptors (Islam et al., 2020).

One to two isoforms of NGFs were identified in all the Asian Cobra venom proteomes with a relative abundance of 0.1 to 6.6% (Table 5.3). Among the venom samples, *N. sumatrana* from Thailand and *N. kaouthia* from north-east India contained relatively higher amounts of NGF at 6.6% and 5.9%, respectively; the rest of the venom samples contained minor amounts (<3%) of NGF (Table 5.3).

5.7.2.4 Cobra Venom Factor

CVFs are high molecular weight (~180 kDa) complement-activating proteins found in Cobra venoms. Usually, they are synthesized as a single pre-pro-protein in the venom glands, and subsequently, they are cleaved into α, β, and γ chains (Alper and Balavitch, 1976; Vogel et al., 1996; Vogel and Fritzinger, 2010). Notably, the α chain of CVF shares significant homology with the human C3 β chain, and the β and γ chains of CVFs resemble C- and N-terminal regions of the C3 α' chain, respectively (Figure 5.19) (Alper and Balavitch, 1976). Due to close sequence similarity with the human C3 protein, CVFs have complement-activating properties, albeit the biological role of CVF in Cobra venoms has never been fully understood. It has been hypothesized that complement activation induced by CVFs at the bite site of victim/prey may promote the release of anaphylatoxins (C3a and C5a) that can enhance local vascular permeability and blood circulation. This effect may aid in faster absorption of other venom toxins into the blood and, subsequently, rapid dissemination of these toxins to their site of action via the blood vascular system (Vogel and Fritzinger, 2010).

With the exceptions of *N. samarensis* from the Philippines, *N. naja* from north-western India, *N. kaouthia* from China, and *N. atra* from Taiwan and Zhejiang, China,

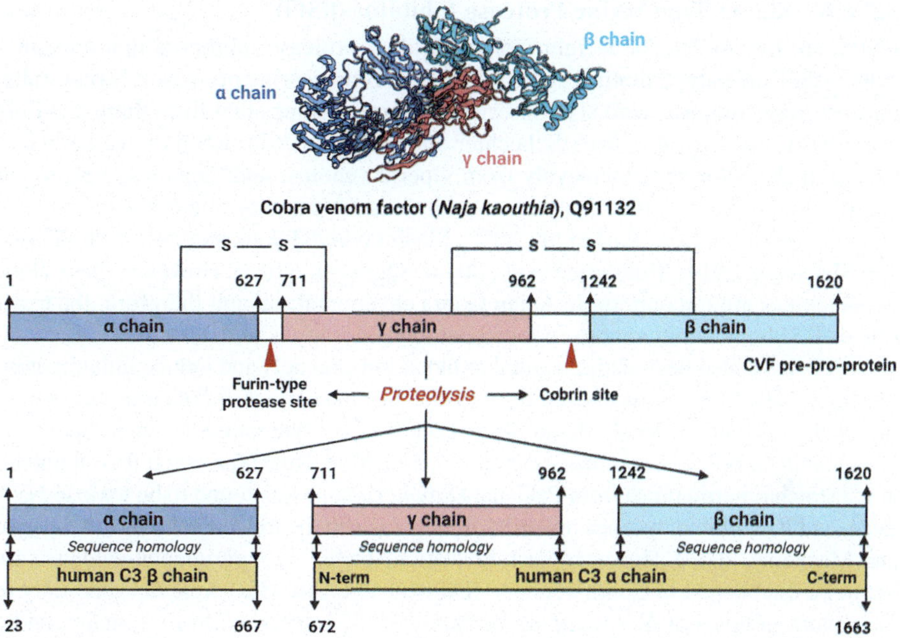

FIGURE 5.19 A schematic diagram showing the primary structure of CVF, its constituent chains (α, β, and γ) that are proteolytically processed as a form of post-translational modification, and sequence homology of the CVF chains to α and β chains of human C3 protein.

the remaining Asian Cobra venom proteomes were reported to contain 1 to 4 isoforms of CVFs with relative abundances ranging from 0.1 to 4.4% (Table 5.3).

5.7.2.5 Vespryn/Ohanin-Like Protein (VES/OLP)

VESs/OLPs (~12 to 24 kDa) are B30.2-like or PRY-SPRY domain-containing proteins primarily reported in Elapidae venoms (except sea snakes) and Viperidae snakes. They have been demonstrated to induce hypernociception, hypolocomotion, and hyperalgesia in experimental mice without any observed lethality, which suggests they are crucial for prey subjugation (Boldrini-Franca et al., 2017; Pung et al., 2005).

Single isoforms of VES/OLP were identified in all Asian Cobra venom proteomes with the exceptions of *N. atra* from China, *N. naja* from north-western India, and *N. sumatrana* from Negeri Sembilan, Malaysia, perhaps due to significantly lower abundances of VES/OLP in these venom samples (Table 5.3). The relative abundances of VES/OLP were reported in the range from 0.1 to 3.8% in most Asian Cobra venom samples; however, three Cobra venom samples (*N. naja* from Karnataka, southern India, *N. sagittifera* from Andaman Island and *N. kaouthia* from China) were reported with unusually higher amounts (6.3 to 9.2%) of VES/OLP (Table 5.3). Although it can be speculated that a higher proportion of VES/OLP may confer superior prey subjugation capabilities to the above Cobra species from these particular geographic locations, there is no experimental evidence to establish such a discrete correlation.

5.7.2.6 Kunitz-Type Serine Protease Inhibitor (KSPI)

KSPIs are low molecular weight (~6 to 10 kDa) protease inhibitors that contain a conserved Kunitz-type motif responsible for inhibiting serine proteases. Structurally, they are very compact, with six conserved cysteine residues that form three disulfide linkages (Calvete et al., 2007; Schechter and Berger, 1967). KSPIs have been isolated and characterized extensively from viperid venoms, and they have a profound contribution to hemostatic disturbances, which is the primary clinical manifestation of viper envenomation (Earl et al., 2012; Mashiko and Takahashi, 2002; Mukherjee and Mackessy, 2014; Mukherjee et al., 2014c; Qiu et al., 2013). However, these clinical outcomes are uncommon in Asian Cobra envenomation, and therefore, the exact role of KSPIs in Cobra venom-induced pathophysiology is undetermined.

KSPIs have also been demonstrated to block ion channels and induce inflammatory responses. Therefore, it may be anticipated that they may affect smooth muscle contraction and contribute to local edema observed in Cobra envenomation. In addition, the Rusvikunin complex (a complex containing two KSPIs, Rusvikunin and Rusvikunin-II at 1:2 stoichiometric ratio) from Pakistan *Daboia russelii* was found to be toxic to NSA mice, and it exhibited dyspnea and hind-limb paresis in the treated animals (Mukherjee and Mackessy, 2014). However, such revelations based on experimental evidence are yet to be established in Cobra venoms. KSPIs (1 to 6 isoforms) were detected only in the venom samples of *N. sagittifera*, *N. sputatrix*, *N. sumatrana* from Sumatra, and a few samples of *N. naja* and *N. kaouthia* with some variations in their relative abundance (0.05 to 5.8%) (Table 5.3).

5.7.2.7 Cystatin (CYS)

CYSs (~10 to 15 kDa) are cysteine protease inhibitors primarily identified in viperid and elapid venoms (Mashiko and Takahashi, 2002; Richards et al., 2011). However, no studies have elucidated their role in venom-induced pathophysiology yet. Single isoforms of CYSs were identified in venoms of only three species of Asian Cobras – *N. sagittifera* (Andaman Cobra), *N. naja* (eastern, western, and southern India), and *N. kaouthia* (north-east India) with relative abundances mostly ranging from 0.01 to 2.1%, thereby suggesting they are minor Cobra venom components. However, the *N. kaouthia* venom from northeast India contained relatively higher proportions of CYSs (8.8%) (Table 5.3).

5.7.2.8 Natriuretic Peptide (NP)

NPs are low molecular weight (~4 to 20 kDa) peptides identified and characterized in elapid and viperid venoms (Suttner and Boldt, 2004; Zhang et al., 2011). A few of these peptides have been demonstrated to promote cGMP production in a dose-dependent manner in endothelial cells, thereby inducing hypotensive effects (Fry et al., 2015; Peterfi et al., 2019; Xu et al., 2015). NPs contribute to Cobra venom-induced pathophysiology by promoting dieresis (kidneys filter too much bodily fluid), natriuresis (excessive loss of sodium in the urine), and vasodilation (Zhang et al., 2011). However, due to their relatively lower abundances in elapid venoms, they have not yet been thoroughly characterized. NPs were identified only in *N. sagittifera* venom, and a few venom samples of *N. naja* (eastern and southern India) and *N. kaouthia* (Thailand and north-east India) at minor proportions (0.1 to 2.9%) (Table 5.3).

5.7.2.9 Other Classes of Non-Enzymatic Toxins

Apart from these non-enzymatic protein families, a few isoforms of VEGF and CTL were identified with very low relative abundances (<0.5%) in a few species of Asian Cobras (Table 5.3). Notably, the above two protein classes are frequently reported in viperid venoms with higher relative abundances (>5–10%), and they have a crucial contribution to the pathophysiology of viper envenomation (Mukherjee et al., 2014a; Yamazaki and Morita, 2006); however, they represent minor venom toxins in Asian Cobra venoms, and their role in Cobra venom-induced pathophysiology is undetermined.

5.8 SPECIES-SPECIFIC AND GEOGRAPHICAL VARIATION IN VENOM COMPOSITION OF ASIAN COBRAS

Depending on the habitat and the availability of certain types of prey, snake venoms have endured strict selection pressures to perform most favorably in a particular habitat (Fry et al., 2008). Although venoms from phylogenetically-related congeneric Asian Cobra species essentially share common trends (for example, an abundance of 3FTxs), subtle yet crucial inter-specific venom variation has also been observed (Chanda et al., 2018; Liu et al., 2017), likely due to differences in their habitat, prey availability, and divergent genetic lineages. Moreover, profound intra-species venom variation owing to disparities in geographical location has also been reported in some species of Asian Cobras (Chanda et al., 2019; Chanda and Mukherjee, 2020b; Chanda et al., 2018; Huang et al., 2015; Tan et al., 2022; Tan et al., 2015).

The differences mentioned above are reflected in protein isoforms and the relative proportion of the identified venom toxins/toxin classes in different venom samples. However, it is noteworthy that using various mass spectrometers, search engines, and protein identification criteria may also contribute to the variations observed. Therefore, comparing findings from independent investigations can be challenging and should be drawn cautiously (Tasoulis and Isbister, 2017). Thus, in the following two sub-sections, we limit our discussion to those proteomic studies performed under parallel or comparable experimental conditions in the same laboratory.

5.8.1 Inter-Species Venom Disparities Among Asian Cobras

In an attempt to analyze the efficacy of Taiwanese bivalent neurotoxic antivenom, Liu et al. (2017) investigated the venom proteomes of *N. atra* (Taiwan), *N. kaouthia* (Thailand), and *N. siamensis* (Thailand) using the venomics approach. The venoms yielded various protein peaks (21 to 36) during the RP-HPLC separation. These peaks were further resolved into 36 to 79 protein bands (Liu et al., 2017). Although 3FTx was the most abundant toxin class (69.0 to 78.3%) in all three venom samples (Figures 5.2d, 5.9d and 5.10), there were distinct differences in the distribution of CTXs and αNTXs. While the venom of *N. atra* was predominated by CTXs (~53%), *N. kaouthia* venom contained more αNTXs (~41%) compared to CTXs (~28%). On the contrary, the relative proportion of CTX and αNTX were nearly comparable (34% vs. 42%) in *N. siamensis* venom (Table 5.3).

The above differences corroborate well with the observed lethality variation (LD_{50} values) among the Cobra species. NTX-dominated *N. kaouthia* and *N. siamensis* venoms have lower LD_{50} values (0.3 to 0.5 μg/g) than CTX-dominated *N. atra* venom (0.6 to 1.0 μg/g). This observation is also consistent with the perception that the relative proportion of αNTXs primarily dictates the lethality of Cobra venoms. Moreover, the relative abundance of PLA_2 in *N. siamensis* and *N. atra* (~15%) was marginally higher compared to that in *N. kaouthia* venom (~12%) (Table 5.3). In addition to the relative distribution of 3FTxs and PLA_2, the authors could also identify specific isoforms specifically expressed in the venom of one or the other species. For example, although the relative abundance of SVMP was found to be comparable among the three Cobra species (~2%), the protein hemorrhagic metalloproteinase-disintegrin-like toxin (named kaouthiagin) was only identified in *N. kaouthia* venom, but not in the other two venom samples. These remarkable and subtle variations underscore the impact of divergent evolution within the same clad of venomous snakes (Liu et al., 2017).

N. naja and *N. kaouthia* are the two major Cobra species distributed on the Indian mainland responsible for many envenoming cases. While *N. naja* is widely distributed across the continent, *N. kaouthia* predominantly inhabits the eastern and north-eastern regions of the country. Although the variation in venom composition of these two species of Cobra from West Bengal, east India, was documented earlier using biochemical and pharmacological characterization (Mukherjee and Maity, 2002), a recent gel-based proteomic investigation shed valuable insight into their subtle yet crucial venom disparities (Figures 5.8h and 5.9a) (Chanda et al., 2018). Although both the venom samples contained comparable amounts of 3FTxs in their proteome, a slightly higher cumulative abundance of CTX and αNTX in *N. naja* venom (~43%) compared to *N. kaouthia* venom (~35%) is likely the cause of higher lethality of the former venom sample (LD_{50} ~0.5 μg/g) than its congener (LD_{50} ~0.7 μg/g). In addition, a marginally higher relative abundance of SVMP and PLA_2 in *N. naja* venom (~26%) compared to *N. kaouthia* venom (22%) might be responsible for higher edema-inducing activity and myotoxicity observed in *N. naja* venom-treated mice. Interestingly, a comparison of the Cobra venom proteomes at individual protein levels suggested the occurrence of 29 common proteins and 23 and 26 uniquely expressed toxins in *N. naja* and *N. kaouthia* venoms, respectively. These observations suggest that it is not only the variation in relative abundance but also the differential expression of distinct toxin isoforms that may drive inter-specific disparities among Cobra venoms, leading to variation in lethality (Chanda et al., 2018).

In a recent proteomic investigation, Attarde et al. (2021) performed an RP-HPLC-based comparative venom profiling of *N. sagittifera* from the Andaman Islands and *N. naja* from southern India. Although a similar number of proteins were identified in these two Cobra species (103 and 101), a closer look into the proteomes revealed exciting differences in the relative abundance of some key toxin sub-classes (Figures 5.3 and 5.8c). In particular, the relative abundance of short-chain αNTXs in *N. sagittifera* (~57%) was significantly higher than that of the *N. naja* venom sample (~30%). On the contrary, the proportion of long-chain αNTX in *N. naja* (~21%) was nearly 40-fold higher than in *N. sagittifera* (~0.5%). The authors proposed that such significant variations in the relative abundance of αNTX sub-classes may have resulted in higher toxicity of *N. sagittifera* venom (LD_{50} ~0.5 μg/g) compared to *N. naja* venom (LD_{50} ~0.8 μg/g) in experimental mice (Attarde et al., 2021). However, a direct correlation between the LD_{50} values of long-chain/short-chain αNTX and their extent of toxicity is yet to be established.

Nevertheless, a few studies have found that short-chain αNTXs isolated from the venoms of *N. kaouthia* and *N. atra* can produce an extended reversible blockage of nAChR compared to long-chain αNTXs (Tan and Tan, 2015). Further, the venom samples of *N. sagittifera* and *N. naja* contained only 2% and 6% of PLA_2 enzymes, respectively. However, the biological and clinical relevance of the occurrence of such unusually lower amounts of PLA_2 in these venom samples is not yet understood. The relative abundance of other major toxin classes, including CTX and CRISP, was comparable among the venoms of *N. sagittifera* and *N. naja* (Attarde et al., 2021).

The venom proteomes of the Asian spitting Cobras *N. samarensis* (southern Philippines), *N. sputatrix* (Java Island), and *N. philippinensis* (south Philippines) were analyzed under comparable conditions in the laboratory of CH Tan using the RP-HPLC-based proteomic approach (Palasuberniam et al., 2021; Tan et al., 2019a; Tan et al., 2017b). A comparison of the findings of these studies revealed remarkable inter-specific differences in the venom composition of these three Asian Cobra species (Figures 5.3–5.6). The proteome of *N. samarensis* was unique and characterized by a predominance of 3FTxs (~90%) and a shallow PLA_2 content (~3%). On the contrary, the venom proteomes of *N. philippinensis* and *N. sputatrix* contained comparable amounts of 3FTxs (~66% and ~64%) and PLA_2 enzymes (~23% and ~31%).

Furthermore, studies have shown profound differences in the distribution of 3FTxs sub-class distribution among the three venoms. In particular, the venoms of *N. samarensis* and *N. philippinensis* were predominated by short-chain αNTX (~66% and ~45%), while the *N. sputatrix* venom sample was abundant in CTXs (~48%). A predominance of αNTXs in the former two venom samples explains the higher lethality of these neurotoxin-driven Asian Cobra venoms (LD_{50} ~0.1 to 0.2 μg/g) compared to the CTX-rich *N. sputatrix* venom (LD_{50} ~0.9 μg/g). It has been observed that the relative abundance of CTXs in the venoms of spitting Cobras is usually higher compared to their non-spitting congeners, and the *N. sputatrix* venom sample with a CTX abundance of ~48% stays true to this particular trend.

However, it is interesting that *N. samarensis* and *N. philippinensis* venoms have evolved with comparatively lower amounts of CTXs (~16% and ~21%) and have retained the venom-spitting trait. This data suggests that perhaps these spitting species deploy other venom toxins, including PLA_2 (~4 to 23%), SVMP (~4 to 5%) and LAAO (~0.2 to 0.4%) for inflicting cytotoxicity and pain-inducing properties (Palasuberniam et al., 2021; Tan et al., 2019a; Tan et al., 2017b). Further studies on the potency of these individual toxins will aid in a better understanding of the extent of cytotoxicity and the relative distribution of these venom toxins in the venoms of Asian spitting Cobras.

5.8.2 Geographical Intra-Species Venom Variability in Asian Cobras

While inter-specific variation in the venom composition of Asian Cobras results from divergent evolution, the intra-species geographical venom variability is primarily driven by subtle differences in ecological niches (Casewell et al., 2020). It is noteworthy that intra-species variation in venom composition has profound implications for differences in severity of clinical presentation and antivenom treatment. In this context, variable LD_{50} values have been reported for venoms of the same species inhabiting distinct geographical locations (Mukherjee and Maity, 1998; Tan et al., 2015).

Furthermore, it has also often been observed that different dosages of antivenoms are required to manage snakebite victims envenomed by the same snake species at other geographic locations. This distinction is primarily because, traditionally, antivenoms are raised against venom samples isolated from the snake species inhabiting a particular geographic area. Therefore, they perform relatively well in these regions. However, the venom composition of a specific species can vary dramatically as we move away from the source location. As a result, the antivenom dosages must be formulated and adapted to the same species of snakes in these distant locations (Kalita et al., 2018a).

One of the first studies that relied on mass spectrometry-based proteomics for elucidating geographical variation in Asian Cobra venoms was reported by the University of Malaysia. Using the venomics approach, the authors analyzed *N. kaouthia* venoms from Malaysia, Thailand, and Vietnam and found striking quantitative differences among the three venom samples (Figure 5.9c–e) (Tan et al., 2015). The study identified 61, 51, and 68 proteins in the *N. kaouthia* venom samples from Malaysia, Thailand, and Vietnam, respectively. Quantitative results indicated that the relative abundance of PLA_2 was the highest in the *N. kaouthia* venom from Malaysia (~23%), followed by Vietnam (~17%). They were the lowest in Thailand's venom sample (~12%).

Furthermore, the venom samples from Thailand (~78%) and Vietnam (~76%) contained relatively higher proportions of 3FTxs in comparison to the Malaysian venom sample (~64%). More importantly, the variable distribution of the sub-classes of 3FTxs among the three venom samples was even more profound. Long-chain αNTXs were predominantly expressed in the *N. kaouthia* venoms from Thailand (~33%), while their relative abundance in the venom sample from Malaysia was significantly lower (~4%). On the other hand, long-chain αNTXs could not be detected in the Vietnamese *N. kaouthia* venom sample, suggesting either they are absent or are present in meagre amounts in this venom sample. The relative abundances of the short-chain αNTXs were also found to vary to some extent – 4.2%, 7.7%, and 9.2% in the *N. kaouthia* venoms from Malaysia, Thailand, and Vietnam, respectively.

Moreover, the venom sample from Vietnam contained unusually higher amounts of neurotoxic WTXs (~19%) compared to those from Malaysia (~9%) and Thailand (~9%). Neurotoxic WTXs act as antagonists of nicotinic receptors, but they are usually less lethal (LD_{50} of ~5 to 80 μg/g) than αNTXs (LD_{50} of ~0.1 μg/g). In addition to the disparities in the relative distribution of αNTXs, the relative proportions of CTXs were also found to vary dramatically. In particular, the venom samples from Malaysia and Vietnam contained significantly higher amounts of CTXs (~45%) than those from Thailand (~28%). These differences have resulted in variable lethalities of the *N. kaouthia* venoms from different locations. With an LD_{50} value of ~0.2 μg/g, the long-chain αNTX-predominated *N. kaouthia* venom from Thailand was found to be the most lethal as compared to the samples from Malaysia and Vietnam (LD_{50} values of ~0.9 μg/g for both venoms) (Tan et al., 2015).

In the same year, Huang et al. (2015) from Taiwan studied the geographic variation in venoms milked from *N. atra* individuals inhabiting eastern and western regions of Taiwan. Although the exact number of proteins identified in both the venom samples was not disclosed, the RP-HPLC-based proteomic investigation indicated some critical differences in the distribution of CTXs and short-chain αNTXs (Figure 5.2b, c). The relative abundance of CTXs in the venom samples from the eastern region was slightly higher (~59%) than that of the western part (~54%). On the contrary, the venom samples milked from the western region contained higher amounts of short-chain αNTXs (~5 to 35%) than those extracted from the western region (~7 to 16%). Nevertheless, the relative abundance of other protein families, including PLA_2, was comparable among the two venom samples. Notably, a higher quantity of short-chain αNTXs in the western Taiwan sample is likely the cause of its higher lethality (LD_{50} value of ~0.3 μg/g) compared to the eastern Taiwan *N. atra* venom (LD_{50} value of ~0.7 μg/g) (Huang et al., 2015).

The geographical variation in the composition of *N. naja* venom from north-western India and Sri Lanka was investigated using the venomics approach (Sintiprungrat et al., 2016). The study identified 25 and 26 proteins in the *N. naja* venom samples from Sri Lanka and north-western India, respectively. Although the relative abundance of 3FTxs in both the venom samples was comparable (~74% vs. ~80%), significant variation was observed in the proportions of αNTXs (Figure 5.8a, b). The relative abundance of this highly potent toxin sub-class was almost double in the Sri Lankan sample (6.2%) compared to the Indian *N. naja* sample (3.7%).

Besides, substantial variation in the PLA_2 content was also detected; the relative abundance of this critical enzymatic class was determined at ~14% and ~21% in *N. naja* venom samples from Sri Lanka and northwestern India, respectively. Nevertheless, the lethality of the venom samples was not determined in the study; therefore, the biological and clinical impact of these disparities is yet to be fully unraveled (Sintiprungrat et al., 2016).

The venom proteomes of *N. naja* from Burdwan (eastern India) and southern India were reported to contain 52 and 45 enzymatic and non-enzymatic toxins (Chanda and Mukherjee, 2020b; Chanda et al., 2018). Although these were independent investigations, they were conducted in the same lab using the gel-based proteomic profiling approach, and therefore, the results can be compared. One of the striking differences between the two venom samples was the predominance of enzymatic toxins in southern India (~47%) compared to eastern India *N. naja* venom (~30%) (Figure 5.8g, h). Furthermore, disparities in the relative distribution of individual toxin classes were also noted.

In particular, the relative abundance of SVMPs was nearly 2.6-fold higher in the venoms of southern India (~16%) compared to eastern India (~6%). On the contrary, the relative abundances of 3FTxs as well as αNTXs were significantly lower in southern India *N. naja* venom (~37% and ~14%, respectively) compared to that of the eastern India counterpart (~61% and ~22%, respectively). The relative abundances of the rest of the protein classes were consistent among the two venom samples. Therefore, the observed lower lethality of the southern India venom sample (LD_{50} ~2.0 μg/g) compared to that of eastern India *N. naja* venom (LD_{50} ~0.7 μg/g) might be because of these critical differences in relative abundances of SVMPs, 3FTxs, and αNTXs in these venom samples.

It is also important to note that an earlier study by Shashidharamurthy et al. (2002) found that at a sub-LD_{50} dose of ~0.5 μg/g, only the venom of eastern India *N. naja* could induce neurotoxic symptoms, while the same effects for the southern India sample could only be observed at higher venom doses (>1.5 μg/g). These disparities in the severity of neurotoxicity are because of the higher relative abundance of αNTXs in the eastern India sample.

Recently, the venom proteomes of Equatorial spitting Cobra *N. sumatrana* from Negeri Sembilan (Malaysia), Penang (Malaysia), Southern Thailand, and Sumatra were analyzed by the RP-HPLC-based proteomic approach that led to the identification of 32, 42, 22 and 55 proteins, respectively (Tan et al., 2022). Apart from the variation in the number of proteins identified, a few quantitative differences among the venom samples were also reported (Figure 5.7a–d). The venoms expressed variable amounts of PLA_2 (~24% to 34%) and 3FTxs (~57% to 70%). Notably, the relative abundances of PLA_2 and CTXs are usually higher in the venoms of spitting Cobras, and this was true in the case of *N. sumatrana*, albeit with marked differences in relative distribution among the venom samples across the geographic locations. The cumulative relative abundances of PLA_2 and CTXs were ~65%, ~76%, ~77%, and ~58% in the venom samples from Negeri Sembilan, Penang, Southern Thailand, and Sumatra, respectively. These findings suggest variable cytotoxicity of these venom samples. However, this difference has neither been experimentally validated nor observed clinically.

Nevertheless, the authors found a strong correlation between the relative abundance of αNTXs and the lethal potency of the above venoms. The αNTX-rich *N. sumatrana* venom samples from Negeri Sembilan, Southern Thailand, and Sumatra (~11% to 16%) exhibited significantly higher lethal potency (LD_{50} ~0.4 to 0.5 μg/g) compared to the venom sample from Penang (LD_{50} ~0.9 μg/g) that contained relatively lower amounts of αNTXs (~5%). In addition, the authors could also notice a few qualitative differences among the venom samples. For example, KSPI was only identified in the venom sample from Sumatra (~6%), SVSPs and CRISPs were not detected in southern Thailand *N. sumatrana* venom, and minor amounts of AChE and 5'-NT (~0.1% to 0.3%) were only identified in the *N. sumatrana* venoms from Penang and Sumatra. However, these are minor components of Asian Cobra venoms and may not directly impact venom-induced lethality (Tan et al., 2022).

The proteome of *N. kaouthia* from Assam, northeast India, was recently deciphered using the gel-based proteomics approach (Kakati et al., 2022). This strategy was employed to investigate the *N. kaouthia* venom of West Bengal, eastern India (Chanda et al., 2018). Furthermore, we can compare the venom proteomes because both studies were performed in the same laboratory. Assam and West Bengal are two neighboring states of India, but they have different geographic landscapes; therefore, some variation in venom composition between the two populations can be anticipated.

A comparison of the two venom proteomes highlighted a few qualitative and quantitative differences (Figure 5.9a, b). The venom proteome of *N. kaouthia* from eastern India was characterized by 55 proteins, while the northeast India venom sample contained only 32 proteins. Some protein classes, such as aminopeptidase (~0.03%) and CYS (~9%), were exclusively identified in north-east India *N. kaouthia* venom. In comparison, hyaluronidases (~0.1%) and some 3FTx sub-classes (MTLP and WTXs) (~23%) were identified only in the venom sample from eastern

India. Apart from these qualitative differences, minor quantitative variations could also be observed for PLA_2s, CRISPs, and NGFs. However, the significant difference between the two venoms was the distribution of αNTXs and CTXs. The relative abundance of CTXs in north-east India *N. kaouthia* venom (~44%) was over twice that of the eastern India venom sample (~20%).

On the contrary, the eastern India venom sample contained slightly higher amounts of αNTXs (~20%) compared to the northeast India counterpart (~14%) (Chanda et al., 2018; Kakati et al., 2022). Although these variations indicate that the two geographically distinct venom samples of *N. kaouthia* can inflict different levels of neurotoxicity and cytotoxicity, such disparities have not yet been experimentally verified.

5.9 CONCLUSION

The venom proteome composition of a snake, including Asian Cobras, determines the resulting pathophysiological effects in envenomed victims. Therefore, a thorough understanding of Asian Cobra venom composition is crucial for comprehending the venom-induced pathophysiological effects and designing efficient antivenoms for treatment. In this context, applying mass spectrometry-based proteomic platforms has significantly advanced our knowledge of Cobra venom composition. In addition, the variation in venom composition owing to inter-specific differences and geographic disparities has also been adequately addressed using proteomics. Nevertheless, proteomic methods face inevitable challenges in identifying novel venom toxins due to their database-dependent nature. Furthermore, due to the unavailability of species-specific Cobra venom protein databases, the quantification could be more precise, and it provides only an estimate of the relative distribution of Cobra venom toxins. Therefore, integrating other omics approaches, including genomics and transcriptomics, may assist in generating species-specific databases, thereby providing an opportunity for an in-depth identification and quantification of Cobra venom toxins.

REFERENCES

Ahmad, S.F., Singchat, W., Panthum, T., Srikulnath, K., 2021. Impact of repetitive DNA elements on snake genome biology and evolution. *Cells* 10, 1707.

Ahmed, M., Rocha, J.B.T., Morsh, V., Schetinger, M.R., Mackessy, S., 2009. Snake venom acetylcholinesterase. *Handbook of venoms and toxins of reptiles*, CRC Press, pp. 207–219.

Alper, C.A., Balavitch, D., 1976. Cobra venom factor: evidence for its being altered cobra C3 (the third component of complement). *Science* 191, 1275–1276.

Antosiewicz, J.M., Shugar, D., 2016. UV-Vis spectroscopy of tyrosine side-groups in studies of protein structure. Part 1: basic principles and properties of tyrosine chromophore. *Biophysical Reviews* 8, 151–161.

Attarde, S., Khochare, S., Iyer, A., Dam, P., Martin, G., Sunagar, K., 2021. Venomics of the enigmatic Andaman cobra (*Naja sagittifera*) and the preclinical failure of Indian antivenoms in Andaman and Nicobar Islands. *Frontiers in Pharmacology* 12, 768210.

Bawaskar, H.S., Bawaskar, P.H., 2004. Envenoming by the common krait (*Bungarus caeruleus*) and Asian cobra (*Naja naja*): Clinical manifestations and their management in a rural setting. *Wilderness & Environmental Medicine* 15, 257–266.

Beraldo Neto, E., Coelho, G.R., Sciani, J.M., Pimenta, D.C., 2021. Proteomic characterization of *Naja mandalayensis* venom. *Journal of Venomous Animals and Toxins Including Tropical Diseases* 27, e20200125.

Boldrini-Franca, J., Cologna, C.T., Pucca, M.B., Bordon, K.C., Amorim, F.G., Anjolette, F.A., Cordeiro, F.A., Wiezel, G.A., Cerni, F.A., Pinheiro-Junior, E.L., Shibao, P.Y., Ferreira, I.G., de Oliveira, I.S., Cardoso, I.A., Arantes, E.C., 2017. Minor snake venom proteins: Structure, function and potential applications. *Biochimica et Biophysica Acta, General Subjects* 1861, 824–838.

Brahma, R.K., McCleary, R.J., Kini, R.M., Doley, R., 2015. Venom gland transcriptomics for identifying, cataloging, and characterizing venom proteins in snakes. *Toxicon* 93, 1–10.

Brown, R.L., Haley, T.L., West, K.A., Crabb, J.W., 1999. Pseudechetoxin: A peptide blocker of cyclic nucleotide-gated ion channels. *Proceedings of the National Academy of Sciences of the United States of America* 96, 754–759.

Burke, J.E., Dennis, E.A., 2009. Phospholipase A2 structure/function, mechanism, and signaling. *Journal of Lipid Research* 50 Suppl, S237–242.

Calvete, J.J., Juarez, P., Sanz, L., 2007. Snake venomics. Strategy and applications. *Journal of Mass Spectrometry : JMS* 42, 1405–1414.

Casewell, N.R., Jackson, T.N.W., Laustsen, A.H., Sunagar, K., 2020. Causes and consequences of snake venom variation. *Trends in Pharmacological Sciences* 41, 570–581.

Casewell, N.R., Wuster, W., Vonk, F.J., Harrison, R.A., Fry, B.G., 2013. Complex cocktails: the evolutionary novelty of venoms. *Trends in Ecology & Evolution* 28, 219–229.

Castoe, T.A., de Koning, A.P., Hall, K.T., Card, D.C., Schield, D.R., Fujita, M.K., Ruggiero, R.P., Degner, J.F., Daza, J.M., Gu, W., Reyes-Velasco, J., Shaney, K.J., Castoe, J.M., Fox, S.E., Poole, A.W., Polanco, D., Dobry, J., Vandewege, M.W., Li, Q., Schott, R.K., Kapusta, A., Minx, P., Feschotte, C., Uetz, P., Ray, D.A., Hoffmann, F.G., Bogden, R., Smith, E.N., Chang, B.S., Vonk, F.J., Casewell, N.R., Henkel, C.V., Richardson, M.K., Mackessy, S.P., Bronikowski, A.M., Yandell, M., Warren, W.C., Secor, S.M., Pollock, D.D., 2013. The Burmese python genome reveals the molecular basis for extreme adaptation in snakes. *Proceedings of the National Academy of Sciences of the United States of America* 110, 20645–20650.

Castoe, T.A., Hall, K.T., Guibotsy Mboulas, M.L., Gu, W., de Koning, A.P., Fox, S.E., Poole, A.W., Vemulapalli, V., Daza, J.M., Mockler, T., Smith, E.N., Feschotte, C., Pollock, D.D., 2011. Discovery of highly divergent repeat landscapes in snake genomes using high-throughput sequencing. *Genome Biology and Evolution* 3, 641–653.

Chanda, A., Kalita, B., Patra, A., Senevirathne, W., Mukherjee, A.K., 2019. Proteomic analysis and antivenomics study of Western India *Naja* venom: correlation between venom composition and clinical manifestations of cobra bite in this region. *Expert Review of Proteomics* 16, 171–184.

Chanda, A., Mukherjee, A.K., 2020a. Mass spectrometric analysis to unravel the venom proteome composition of Indian snakes: opening new avenues in clinical research. *Expert Review of Proteomics* 17, 411–423.

Chanda, A., Mukherjee, A.K., 2020b. Quantitative proteomics to reveal the composition of Southern India spectacled cobra (*Naja naja*) venom and its immunological cross-reactivity towards commercial antivenom. *International Journal of Biological Macromolecules* 160, 224–232.

Chanda, A., Patra, A., Kalita, B., Mukherjee, A.K., 2018. Proteomics analysis to compare the venom composition between *Naja* and *Naja kaouthia* from the same geographical location of eastern India: Correlation with pathophysiology of envenomation and immunological cross-reactivity towards commercial polyantivenom. *Expert Review of Proteomics* 15, 949–961.

Chapeaurouge, A., Reza, M.A., Mackessy, S.P., Carvalho, P.C., Valente, R.H., Teixeira-Ferreira, A., Perales, J., Lin, Q., Kini, R.M., 2015. Interrogating the venom of the viperid snake *Sistrurus catenatus edwardsii* by a combined approach of electrospray and MALDI mass spectrometry. *PLoS One* 10, e0092091.

Choi, H., Fermin, D., Nesvizhskii, A.I., 2008. Significance analysis of spectral count data in label-free shotgun proteomics. *Molecular & Cellular Proteomics : MCP* 7, 2373–2385.

Chong, H.P., Tan, K.Y., Tan, N.H., Tan, C.H., 2019. Exploring the diversity and novelty of toxin genes in *Naja sumatrana*, the equatorial spitting cobra from Malaysia THROUGH De Novo venom-gland transcriptomics. *Toxins* 11, 104.

Daltry, J.C., Wuster, W., Thorpe, R.S., 1996. Diet and snake venom evolution. *Nature* 379, 537–540.

Deka, A., Bhatia, S., Santra, V., Bharti, O.K., Lalremsanga, H.T., Martin, G., Wuster, W., Owens, J.B., Graham, S., Doley, R., Malhotra, A., 2023. Multilevel comparison of Indian *Naja* venoms and their cross-reactivity with Indian polyvalent antivenoms. *Toxins* 15, 258.

Dennis, E.A., Cao, J., Hsu, Y.H., Magrioti, V., Kokotos, G., 2011. Phospholipase A2 enzymes: physical structure, biological function, disease implication, chemical inhibition, and therapeutic intervention. *Chemical Reviews* 111, 6130–6185.

Dhananjaya, B.L., D'Souza, C.J., 2010. The pharmacological role of nucleotidases in snake venoms. *Cell Biochemistry and Function* 28, 171–177.

Doley, R., King, G.F., Mukherjee, A.K., 2004. Differential hydrolysis of erythrocyte and mitochondrial membrane phospholipids by two phospholipase A2 isoenzymes (NK-PLA2-I and NK-PLA2-II) from the venom of the Indian monocled cobra *Naja kaouthia*. *Archives of Biochemistry and Biophysics* 425, 1–13.

Doley, R., Mukherjee, A.K., 2003. Purification and characterization of an anticoagulant phospholipase A(2) from Indian monocled cobra (*Naja kaouthia*) venom. *Toxicon* 41, 81–91.

Du, X.Y., Clemetson, K.J., 2002. Snake venom L-amino acid oxidases. *Toxicon* 40, 659–665.

Dufton, M.J., Hider, R.C., 1983. Conformational properties of the neurotoxins and cytotoxins isolated from Elapid snake venoms. *CRC Critical Reviews in Biochemistry* 14, 113–171.

Dutta, S., Chanda, A., Kalita, B., Islam, T., Patra, A., Mukherjee, A.K., 2017. Proteomic analysis to unravel the complex venom proteome of eastern India *Naja naja*: Correlation of venom composition with its biochemical and pharmacological properties. *Journal of Proteomics* 156, 29–39.

Dutta, S., Gogoi, D., Mukherjee, A.K., 2015. Anticoagulant mechanism and platelet deaggregation property of a non-cytotoxic, acidic phospholipase A2 purified from Indian cobra (*Naja naja*) venom: inhibition of anticoagulant activity by low molecular weight heparin. *Biochimie* 110, 93–106.

Dutta, S., Sinha, A., Dasgupta, S., Mukherjee, A.K., 2019. Binding of a *Naja naja* venom acidic phospholipase A(2) cognate complex to membrane-bound vimentin of rat L6 cells: Implications in cobra venom-induced cytotoxicity. *Biochimica et Biophysica Acta, Biomembranes* 1861, 958–977.

Earl, S.T., Richards, R., Johnson, L.A., Flight, S., Anderson, S., Liao, A., de Jersey, J., Masci, P.P., Lavin, M.F., 2012. Identification and characterisation of Kunitz-type plasma kallikrein inhibitors unique to *Oxyuranus* sp. snake venoms. *Biochimie* 94, 365–373.

Fox, J.W., 2013. A brief review of the scientific history of several lesser-known snake venom proteins: l-amino acid oxidases, hyaluronidases and phosphodiesterases. *Toxicon* 62, 75–82.

Fox, J.W., Serrano, S.M., 2008. Insights into and speculations about snake venom metalloproteinase (SVMP) synthesis, folding and disulfide bond formation and their contribution to venom complexity. *The FEBS Journal* 275, 3016–3030.

Frobert, Y., Creminon, C., Cousin, X., Remy, M.H., Chatel, J.M., Bon, S., Bon, C., Grassi, J., 1997. Acetylcholinesterases from Elapidae snake venoms: Biochemical, immunological and enzymatic characterization. *Biochimica et Biophysica Acta* 1339, 253–267.

Fry, B., Jackson, T., Takacs, Z., Reeks, T., Sunagar, K., 2015. C-type natriuretic peptides. Venomous reptiles and their toxins: Evolution, pathophysiology and biodiscovery. Oxford University Press, pp. 318–326.

Fry, B.G., 2005. From genome to "venome": molecular origin and evolution of the snake venom proteome inferred from phylogenetic analysis of toxin sequences and related body proteins. *Genome Research* 15, 403–420.

Fry, B.G., Scheib, H., van der Weerd, L., Young, B., McNaughtan, J., Ramjan, S.F., Vidal, N., Poelmann, R.E., Norman, J.A., 2008. Evolution of an arsenal: structural and functional diversification of the venom system in the advanced snakes (Caenophidia). *Molecular & Cellular Proteomics : MCP* 7, 215–246.

Fry, B.G., Wuster, W., Kini, R.M., Brusic, V., Khan, A., Venkataraman, D., Rooney, A.P., 2003. Molecular evolution and phylogeny of elapid snake venom three-finger toxins. *Journal of Molecular Evolution* 57, 110–129.

Georgieva, D., Risch, M., Kardas, A., Buck, F., von Bergen, M., Betzel, C., 2008. Comparative analysis of the venom proteomes of *Vipera ammodytes ammodytes* and *Vipera ammodytes meridionalis*. *Journal of Proteome Research* 7, 866–886.

Gutierrez, J.M., Calvete, J.J., Habib, A.G., Harrison, R.A., Williams, D.J., Warrell, D.A., 2017. *Snakebite envenoming*. *Nature Reviews. Disease Primers* 3, 17079.

Gutierrez, J.M., Rucavado, A., 2000. Snake venom metalloproteinases: their role in the pathogenesis of local tissue damage. *Biochimie* 82, 841–850.

Hamako, J., Matsui, T., Nishida, S., Nomura, S., Fujimura, Y., Ito, M., Ozeki, Y., Titani, K., 1998. Purification and characterization of kaouthiagin, a von Willebrand factor-binding and -cleaving metalloproteinase from *Naja kaouthia* cobra venom. *Thrombosis and Haemostasis* 80, 499–505.

Hiu, J.J., Yap, M.K.K., 2020. Cytotoxicity of snake venom enzymatic toxins: phospholipase A2 and l-amino acid oxidase. *Biochemical Society Transactions* 48, 719–731.

Holding, M.L., Margres, M.J., Mason, A.J., Parkinson, C.L., Rokyta, D.R., 2018. Evaluating the performance of De Novo assembly methods for venom-gland transcriptomics. *Toxins* 10, 249.

Huang, H.W., Liu, B.S., Chien, K.Y., Chiang, L.C., Huang, S.Y., Sung, W.C., Wu, W.G., 2015. Cobra venom proteome and glycome determined from individual snakes of *Naja atra* reveal medically important dynamic range and systematic geographic variation. *Journal of Proteomics* 128, 92–104.

Islam, T., Madhubala, D., Mukhopadhyay, R., Mukherjee, A.K., 2021. Transcriptomic and functional proteomics analyses to unveil the common and unique pathway(s) of neuritogenesis induced by Russell's viper venom nerve growth factor in rat pheochromocytoma neuronal cells. *Expert Review of Proteomics* 18, 463–481.

Islam, T., Majumder, M., Bidkar, A., Ghosh, S.S., Mukhopadhyay, R., Utkin, Y., Mukherjee, A.K., 2020. Nerve growth factor from Indian Russell's viper venom (RVV-NGFa) shows high affinity binding to TrkA receptor expressed in breast cancer cells: Application of fluorescence labeled RVV-NGFa in the clinical diagnosis of breast cancer. *Biochimie* 176, 31–44.

Jagadeesha, D.K., Shashidhara Murthy, R., Girish, K.S., Kemparaju, K., 2002. A non-toxic anticoagulant metalloprotease: purification and characterization from Indian cobra (*Naja naja naja*) venom. *Toxicon* 40, 667–675.

Kakati, H., Patra, A., Kalita, B., Chanda, A., Rapole, S., Mukherjee, A.K., 2022. A comparison of two different analytical workflows to determine the venom proteome composition of *Naja kaouthia* from North-East India and immunological profiling of venom against commercial antivenoms. *International Journal of Biological Macromolecules* 208, 275–287.

Kalita, B., Mackessy, S.P., Mukherjee, A.K., 2018a. Proteomic analysis reveals geographic variation in venom composition of Russell's viper in the Indian subcontinent: implications for clinical manifestations post-envenomation and antivenom treatment. *Expert Review of Proteomics* 15, 837–849.

Kalita, B., Mukherjee, A.K., 2019. Recent advances in snake venom proteomics research in India: A new horizon to decipher the geographical variation in venom proteome composition and exploration of candidate drug prototypes. *Journal of Proteins and Proteomics* 10, 149–164.

Kalita, B., Patra, A., Das, A., Mukherjee, A.K., 2018b. Proteomic analysis and immuno-profiling of Eastern India Russell's viper (*Daboia russelii*) venom: Correlation between RVV composition and clinical manifestations post RV bite. *Journal of Proteome Research* 17, 2819–2833.

Kalita, B., Patra, A., Jahan, S., Mukherjee, A.K., 2018c. First report of the characterization of a snake venom apyrase (Ruviapyrase) from Indian Russell's viper (*Daboia russelii*) venom. *International Journal of Biological Macromolecules* 111, 639–648.

Kalita, B., Patra, A., Mukherjee, A.K., 2017. Unraveling the proteome composition and immuno-profiling of Western India Russell's viper venom for in-depth understanding of its pharmacological properties, clinical manifestations, and effective antivenom treatment. *Journal of Proteome Research* 16, 583–598.

Kazandjian, T.D., Petras, D., Robinson, S.D., van Thiel, J., Greene, H.W., Arbuckle, K., Barlow, A., Carter, D.A., Wouters, R.M., Whiteley, G., Wagstaff, S.C., Arias, A.S., Albulescu, L.O., Plettenberg Laing, A., Hall, C., Heap, A., Penrhyn-Lowe, S., McCabe, C.V., Ainsworth, S., da Silva, R.R., Dorrestein, P.C., Richardson, M.K., Gutierrez, J.M., Calvete, J.J., Harrison, R.A., Vetter, I., Undheim, E.A.B., Wuster, W., Casewell, N.R., 2021. Convergent evolution of pain-inducing defensive venom components in spitting cobras. *Science* 371, 386–390.

Kini, R.M., 2006. Serine proteases affecting blood coagulation and fibrinolysis from snake venoms. *Pathophysiology of Haemostasis and Thrombosis* 34, 200–204.

Kini, R.M., Doley, R., 2010. Structure, function and evolution of three-finger toxins: Mini proteins with multiple targets. *Toxicon* 56, 855–867.

Koludarov, I., Senoner, T., Jackson, T.N.W., Dashevsky, D., Heinzinger, M., Aird, S.D., Rost, B., 2023. Domain loss enabled evolution of novel functions in the snake three-finger toxin gene superfamily. *Nature Communications* 14, 4861.

Kostiza, T., Meier, J., 1996. Nerve growth factors from snake venoms: Chemical properties, mode of action and biological significance. *Toxicon* 34, 787–806.

Kularatne, S.A., Budagoda, B.D., Gawarammana, I.B., Kularatne, W.K., 2009. Epidemiology, clinical profile and management issues of cobra (*Naja naja*) bites in Sri Lanka: First authenticated case series. *Transactions of the Royal Society of Tropical Medicine and Hygiene* 103, 924–930.

Kulkeaw, K., Chaicumpa, W., Sakolvaree, Y., Tongtawe, P., Tapchaisri, P., 2007. Proteome and immunome of the venom of the Thai cobra, *Naja kaouthia*. *Toxicon* 49, 1026–1041.

Kumar, M.S., Devaraj, V.R., Vishwanath, B.S., Kemparaju, K., 2010. Anti-coagulant activity of a metalloprotease: Further characterization from the Indian cobra (*Naja naja*) venom. *Journal of Thrombosis and Thrombolysis* 29, 340–348.

Laustsen, A.H., Gutierrez, J.M., Lohse, B., Rasmussen, A.R., Fernandez, J., Milbo, C., Lomonte, B., 2015. Snake venomics of monocled cobra (*Naja kaouthia*) and investigation of human IgG response against venom toxins. *Toxicon* 99, 23–35.

Liu, C.C., You, C.H., Wang, P.J., Yu, J.S., Huang, G.J., Liu, C.H., Hsieh, W.C., Lin, C.C., 2017. Analysis of the efficacy of Taiwanese freeze-dried neurotoxic antivenom against *Naja kaouthia*, *Naja siamensis* and *Ophiophagus hannah* through proteomics and animal model approaches. *PLoS Neglected Tropical Diseases* 11, e0006138.

Lomonte, B., Calvete, J.J., 2017. Strategies in 'snake venomics' aiming at an integrative view of compositional, functional, and immunological characteristics of venoms. *The Journal of Venomous Animals and Toxins Including Tropical Diseases* 23, 26.

Mackessy, S.P., 2002. Biochemistry and pharmacology of colubrid snake venoms. *Journal of Toxicology - Toxin Reviews* 21, 43–83.

Mackessy, S.P., 2010. Evolutionary trends in venom composition in the western rattlesnakes (*Crotalus viridis* sensu lato): toxicity vs. tenderizers. *Toxicon* 55, 1463–1474.

Manuwar, A., Dreyer, B., Bohmert, A., Ullah, A., Mughal, Z., Akrem, A., Ali, S.A., Schluter, H., Betzel, C., 2020. Proteomic investigations of two Pakistani *Naja* snake venoms species unravel the venom complexity, posttranslational modifications, and presence of extracellular vesicles. *Toxins* 12, 669.

Markland, F.S., Jr., 1998. Snake venom fibrinogenolytic and fibrinolytic enzymes: An updated inventory. Registry of exogenous hemostatic factors of the scientific and standardization committee of the international society on thrombosis and haemostasis. *Thrombosis and Haemostasis* 79, 668–674.

Mashiko, H., Takahashi, H., 2002. Cysteine proteinase inhibitors in elapid and hydrophiid snake venoms. *Toxicon* 40, 1275–1281.
Matsui, T., Fujimura, Y., Titani, K., 2000. Snake venom proteases affecting hemostasis and thrombosis. *Biochimica et Biophysica Acta* 1477, 146–156.
Modahl, C.M., Mukherjee, A.K., Mackessy, S.P., 2016. An analysis of venom ontogeny and prey-specific toxicity in the monocled cobra (*Naja kaouthia*). *Toxicon* 119, 8–20.
More, S.S., Kiran, K., Veena, S., Gadag, J., 2010. Purification of an L-amino acid oxidase from *Bungarus caeruleus* (Indian krait) venom. *Journal of Venomous Animals and Toxins Including Tropical Diseases* 16, 60–76.
Mukherjee, A.K., 2008. Characterization of a novel pro-coagulant metalloprotease (RVBCMP) possessing alpha-fibrinogenase and tissue haemorrhagic activity from venom of *Daboia russelli russelli* (Russell's viper): evidence of distinct coagulant and haemorrhagic sites in RVBCMP. *Toxicon* 51, 923–933.
Mukherjee, A.K., 2014a. A major phospholipase A2 from *Daboia russelii russelii* venom shows potent anticoagulant action via thrombin inhibition and binding with plasma phospholipids. *Biochimie* 99, 153–161.
Mukherjee, A.K., 2014b. The pro-coagulant fibrinogenolytic serine protease isoenzymes purified from *Daboia russelii russelii* venom coagulate the blood through factor V activation: role of glycosylation on enzymatic activity. *PLoS One* 9, e86823.
Mukherjee, A.K., Dutta, S., Mackessy, S.P., 2014a. A new C-type lectin (RVsnaclec) purified from venom of *Daboia russelii russelii* shows anticoagulant activity via inhibition of FXa and concentration-dependent differential response to platelets in a Ca(2)(+)-independent manner. *Thrombosis Research* 134, 1150–1156.
Mukherjee, A.K., Kalita, B., Mackessy, S.P., 2016. A proteomic analysis of Pakistan *Daboia russelii russelii* venom and assessment of potency of Indian polyvalent and monovalent antivenom. *Journal of Proteomics* 144, 73–86.
Mukherjee, A.K., Kalita, B., Thakur, R., 2014b. Two acidic, anticoagulant PLA2 isoenzymes purified from the venom of monocled cobra *Naja kaouthia* exhibit different potency to inhibit thrombin and factor Xa via phospholipids independent, non-enzymatic mechanism. *PLoS One* 9, e101334.
Mukherjee, A.K., Mackessy, S.P., 2013. Biochemical and pharmacological properties of a new thrombin-like serine protease (Russelobin) from the venom of Russell's viper (*Daboia russelii russelii*) and assessment of its therapeutic potential. *Biochimica et Biophysica Acta* 1830, 3476–3488.
Mukherjee, A.K., Mackessy, S.P., 2014. Pharmacological properties and pathophysiological significance of a Kunitz-type protease inhibitor (Rusvikunin-II) and its protein complex (Rusvikunin complex) purified from *Daboia russelii russelii* venom. *Toxicon* 89, 55–66.
Mukherjee, A.K., Mackessy, S.P., Dutta, S., 2014c. Characterization of a Kunitz-type protease inhibitor peptide (Rusvikunin) purified from *Daboia russelii russelii* venom. *International Journal of Biological Macromolecules* 67, 154–162.
Mukherjee, A.K., Maity, C.R., 1998. The composition of *Naja naja* venom samples from three districts of West Bengal, India. Comparative biochemistry and physiology. *Comparative Biochemistry and Physioloy. Part A, Molecular & Integrative Physiology* 119, 621–627.
Mukherjee, A.K., Maity, C.R., 2002. Biochemical composition, lethality and pathophysiology of venom from two cobras—*Naja naja* and *N. kaouthia*. Comparative biochemistry and physiology. *Comparative Biochemistry and Physiology. Part B, Biochemistry & Molecular Biology* 131, 125–132.
Mukherjee, A.K., Saviola, A.J., Burns, P.D., Mackessy, S.P., 2015. Apoptosis induction in human breast cancer (MCF-7) cells by a novel venom L-amino acid oxidase (Rusvinoxidase) is independent of its enzymatic activity and is accompanied by caspase-7 activation and reactive oxygen species production. *Apoptosis: An International Journal on Programmed Cell Death* 20, 1358–1372.

Naumann, G.B., Silva, L.F., Silva, L., Faria, G., Richardson, M., Evangelista, K., Kohlhoff, M., Gontijo, C.M., Navdaev, A., de Rezende, F.F., Eble, J.A., Sanchez, E.F., 2011. Cytotoxicity and inhibition of platelet aggregation caused by an l-amino acid oxidase from *Bothrops leucurus* venom. *Biochimica et Biophysica Acta* 1810, 683–694.

Nesvizhskii, A.I., Aebersold, R., 2005. Interpretation of shotgun proteomic data: The protein inference problem. *Molecular & Cellular Proteomics : MCP* 4, 1419–1440.

Neuhoff, V., Stamm, R., Eibl, H., 1985. Clear background and highly sensitive protein staining with coomassie blue dyes in polyacrylamide gels: A systematic analysis. *Electrophoresis* 6, 427–448.

Ogawa, Y., Murayama, N., Yanoshita, R., 2009. Molecular cloning and characterization of ecto-5'-nucleotidase from the venoms of *Gloydius blomhoffi*. *Toxicon* 54, 408–412.

Ohno, M., Menez, R., Ogawa, T., Danse, J., Shimohigashi, Y., Fromen, C., Ducancel, F., Zinn-Justin, S., Boulain, J.-C., Tamiya, T., 1997. Molecular evolution of snake toxins: is the functional diversify of snake toxins associated with a mechanism of accelerated evolution? *Progress in Nucleic Acid Research and Molecular Biology* 59, 307–364.

Ohno, S., 2013. Evolution by gene duplication. Springer Science & Business Media.

Oliveira, A.K., Paes Leme, A.F., Asega, A.F., Camargo, A.C., Fox, J.W., Serrano, S.M., 2010. New insights into the structural elements involved in the skin haemorrhage induced by snake venom metalloproteinases. *Thrombosis and Haemostasis* 104, 485–497.

Oliveira, I.S., Pucca, M.B., Ferreira, I.G., Cerni, F.A., Jacob, B., Wiezel, G.A., Pinheiro-Junior, E.L., Cordeiro, F.A., Bordon, K.C.F., Arantes, E.C., 2022. State-of-the-art review of snake venom phosphodiesterases (svPDEs). *Toxicon* 217, 121–130.

Osipov, A.V., Levashov, M.Y., Tsetlin, V.I., Utkin, Y.N., 2005. Cobra venom contains a pool of cysteine-rich secretory proteins. *Biochemical and Biophysical Research Communications* 328, 177–182.

Osset, M., Pinol, M., Fallon, M.J., de Llorens, R., Cuchillo, C.M., 1989. Interference of the carbohydrate moiety in coomassie brilliant blue R-250 protein staining. *Electrophoresis* 10, 271–273.

Palasuberniam, P., Chan, Y.W., Tan, K.Y., Tan, C.H., 2021. Snake venom proteomics of samar cobra (*Naja samarensis*) from the Southern Philippines: Short alpha-neurotoxins as the dominant lethal component weakly cross-neutralized by the Philippine cobra antivenom. *Frontiers in Pharmacology* 12, 727756.

Panagides, N., Jackson, T.N., Ikonomopoulou, M.P., Arbuckle, K., Pretzler, R., Yang, D.C., Ali, S.A., Koludarov, I., Dobson, J., Sanker, B., Asselin, A., Santana, R.C., Hendrikx, I., van der Ploeg, H., Tai, A.P.J., van den Bergh, R., Kerkkamp, H.M., Vonk, F.J., Naude, A., Strydom, M.A., Jacobsz, L., Dunstan, N., Jaeger, M., Hodgson, W.C., Miles, J., Fry, B.G., 2017. How the cobra got its flesh-eating venom: Cytotoxicity as a defensive innovation and its co-evolution with hooding, aposematic marking, and spitting. *Toxins* 9, 103.

Pasquesi, G.I.M., Adams, R.H., Card, D.C., Schield, D.R., Corbin, A.B., Perry, B.W., Reyes-Velasco, J., Ruggiero, R.P., Vandewege, M.W., Shortt, J.A., Castoe, T.A., 2018. Squamate reptiles challenge paradigms of genomic repeat element evolution set by birds and mammals. *Nature Communications* 9, 2774.

Patra, A., Chanda, A., Mukherjee, A.K., 2019. Quantitative proteomic analysis of venom from Southern India common krait (*Bungarus caeruleus*) and identification of poorly immunogenic toxins by immune-profiling against commercial antivenom. *Expert Review of Proteomics* 16, 457–469.

Patra, A., Kalita, B., Chanda, A., Mukherjee, A.K., 2017. Proteomics and antivenomics of *Echis carinatus carinatus* venom: Correlation with pharmacological properties and pathophysiology of envenomation. *Scientific Reports* 7, 17119.

Peterfi, O., Boda, F., Szabo, Z., Ferencz, E., Baba, L., 2019. Hypotensive snake venom components—A mini-review. *Molecules* 24, 2778.

Pung, Y.F., Wong, P.T., Kumar, P.P., Hodgson, W.C., Kini, R.M., 2005. Ohanin, a novel protein from king cobra venom, induces hypolocomotion and hyperalgesia in mice. *The Journal of Biological Chemistry* 280, 13137–13147.

Qiu, Y., Lee, K.S., Choo, Y.M., Kong, D., Yoon, H.J., Jin, B.R., 2013. Molecular cloning and antifibrinolytic activity of a serine protease inhibitor from bumblebee (*Bombus terrestris*) venom. *Toxicon* 63, 1–6.

Raba, R., Aaviksaar, A., Raba, M., Siigur, J., 1979. Cobra venom acetylcholinesterase. Purification and molecular properties. *European Journal of Biochemistry* 96, 151–158.

Richards, R., St Pierre, L., Trabi, M., Johnson, L.A., de Jersey, J., Masci, P.P., Lavin, M.F., 2011. Cloning and characterisation of novel cystatins from elapid snake venom glands. *Biochimie* 93, 659–668.

Ryabinin, V., Ziganshin, R., Starkov, V., Tsetlin, V., Utkin, Y., 2019. Intraspecific variability in the composition of the venom from monocled cobra (*Naja kaouthia*). *Russian Journal of Bioorganic Chemistry* 45, 107–121.

Saikia, D., Bordoloi, N.K., Chattopadhyay, P., Choklingam, S., Ghosh, S.S., Mukherjee, A.K., 2012. Differential mode of attack on membrane phospholipids by an acidic phospholipase A(2) (RVVA-PLA(2)-I) from *Daboia russelli* venom. *Biochimica et Biophysica Acta* 1818, 3149–3157.

Sanchez, E.F., Flores-Ortiz, R.J., Alvarenga, V.G., Eble, J.A., 2017. Direct fibrinolytic snake venom metalloproteinases affecting hemostasis: Structural, biochemical features and therapeutic potential. *Toxins* 9, 392.

Santoro, M.L., Vaquero, T.S., Paes Leme, A.F., Serrano, S.M., 2009. NPP-BJ, a nucleotide pyrophosphatase/phosphodiesterase from *Bothrops jararaca* snake venom, inhibits platelet aggregation. *Toxicon* 54, 499–512.

Schechter, I., Berger, A., 1967. On the size of the active site in proteases. I. Papain. *Biochemical and Biophysical Research Communications* 27, 157–162.

Schield, D.R., Card, D.C., Hales, N.R., Perry, B.W., Pasquesi, G.M., Blackmon, H., Adams, R.H., Corbin, A.B., Smith, C.F., Ramesh, B., Demuth, J.P., Betran, E., Tollis, M., Meik, J.M., Mackessy, S.P., Castoe, T.A., 2019. The origins and evolution of chromosomes, dosage compensation, and mechanisms underlying venom regulation in snakes. *Genome Research* 29, 590–601.

Shan, L.L., Gao, J.F., Zhang, Y.X., Shen, S.S., He, Y., Wang, J., Ma, X.M., Ji, X., 2016. Proteomic characterization and comparison of venoms from two elapid snakes (*Bungarus multicinctus* and *Naja atra*) from China. *Journal of Proteomics* 138, 83–94.

Shashidharamurthy, R., Jagadeesha, D.K., Girish, K.S., Kemparaju, K., 2002. Variations in biochemical and pharmacological properties of Indian cobra (*Naja naja naja*) venom due to geographical distribution. *Molecular and Cellular Biochemistry* 229, 93–101.

Shibata, H., Chijiwa, T., Oda-Ueda, N., Nakamura, H., Yamaguchi, K., Hattori, S., Matsubara, K., Matsuda, Y., Yamashita, A., Isomoto, A., Mori, K., Tashiro, K., Kuhara, S., Yamasaki, S., Fujie, M., Goto, H., Koyanagi, R., Takeuchi, T., Fukumaki, Y., Ohno, M., Shoguchi, E., Hisata, K., Satoh, N., Ogawa, T., 2018. The habu genome reveals accelerated evolution of venom protein genes. *Scientific Reports* 8, 11300.

Siigur, E., Aaspollu, A., Siigur, J., 2003. Anticoagulant serine fibrinogenases from *Vipera lebetina* venom: structure-function relationships. *Thrombosis and Haemostasis* 89, 826–831.

Sintiprungrat, K., Watcharatanyatip, K., Senevirathne, W.D., Chaisuriya, P., Chokchaichamnankit, D., Srisomsap, C., Ratanabanangkoon, K., 2016. A comparative study of venomics of *Naja naja* from India and Sri Lanka, clinical manifestations and antivenomics of an Indian polyspecific antivenom. *Journal of Proteomics* 132, 131–143.

Slotta, K.H., Vick, J.A., 1969. Identification of the direct lytic factor from cobra venom as cardiotoxin. *Toxicon* 6, 167–173.

Smith, C.F., Mackessy, S.P., 2020. Asymmetrical expression of toxins between the left and right venom glands of an individual prairie rattlesnake (*Crotalus viridis viridis*). *Toxicon* 186, 105–108.

Sun, Q.Y., Bao, J., 2010. Purification, cloning and characterization of a metalloproteinase from *Naja atra* venom. *Toxicon* 56, 1459–1469.

Sun, Q.Y., Wang, C.E., Li, Y.N., Bao, J., 2020. Inhibition of platelet aggregation and blood coagulation by a P-III class metalloproteinase purified from *Naja atra* venom. *Toxicon* 187, 223–231.

Sunagar, K., Morgenstern, D., Reitzel, A.M., Moran, Y., 2016. Ecological venomics: How genomics, transcriptomics and proteomics can shed new light on the ecology and evolution of venom. *Journal of Proteomics* 135, 62–72.

Suryamohan, K., Krishnankutty, S.P., Guillory, J., Jevit, M., Schroder, M.S., Wu, M., Kuriakose, B., Mathew, O.K., Perumal, R.C., Koludarov, I., Goldstein, L.D., Senger, K., Dixon, M.D., Velayutham, D., Vargas, D., Chaudhuri, S., Muraleedharan, M., Goel, R., Chen, Y.J., Ratan, A., Liu, P., Faherty, B., de la Rosa, G., Shibata, H., Baca, M., Sagolla, M., Ziai, J., Wright, G.A., Vucic, D., Mohan, S., Antony, A., Stinson, J., Kirkpatrick, D.S., Hannoush, R.N., Durinck, S., Modrusan, Z., Stawiski, E.W., Wiley, K., Raudsepp, T., Kini, R.M., Zachariah, A., Seshagiri, S., 2020. The Indian cobra reference genome and transcriptome enables comprehensive identification of venom toxins. *Nature Genetics* 52, 106–117.

Suttner, S.W., Boldt, J., 2004. Natriuretic peptide system: physiology and clinical utility. *Current Opinion in Critical Care* 10, 336–341.

Tadokoro, T., Modahl, C.M., Maenaka, K., Aoki-Shioi, N., 2020. Cysteine-rich secretory proteins (CRISPs) from venomous snakes: An overview of the functional diversity in a large and underappreciated superfamily. *Toxins* 12, 175.

Tan, C.H., Tan, K.Y., Wong, K.Y., Tan, N.H., Chong, H.P., 2022. Equatorial spitting cobra (*Naja sumatrana*) from Malaysia (Negeri Sembilan and Penang), Southern Thailand, and Sumatra: Comparative venom proteomics, immunoreactivity and cross-neutralization by antivenom. *Toxins* 14, 522.

Tan, C.H., Tan, N.H., 2015. Toxinology of snake venoms: The Malaysian context. *Snake Venoms*, 1–37.

Tan, C.H., Wong, K.Y., Chong, H.P., Tan, N.H., Tan, K.Y., 2019a. Proteomic insights into short neurotoxin-driven, highly neurotoxic venom of Philippine cobra (*Naja philippinensis*) and toxicity correlation of cobra envenomation in Asia. *Journal of Proteomics* 206, 103418.

Tan, C.H., Wong, K.Y., Tan, N.H., Ng, T.S., Tan, K.Y., 2019b. Distinctive distribution of secretory phospholipases A(2) in the venoms of Afro-Asian cobras (Subgenus: Naja, Afronaja, Boulengerina and Uraeus). *Toxins* 11, 116.

Tan, K.Y., Tan, C.H., Chanhome, L., Tan, N.H., 2017a. Comparative venom gland transcriptomics of *Naja kaouthia* (monocled cobra) from Malaysia and Thailand: elucidating geographical venom variation and insights into sequence novelty. *PeerJ* 5, e3142.

Tan, K.Y., Tan, C.H., Fung, S.Y., Tan, N.H., 2015. Venomics, lethality and neutralization of *Naja kaouthia* (monocled cobra) venoms from three different geographical regions of Southeast Asia. *Journal of Proteomics* 120, 105–125.

Tan, N.H., Wong, K.Y., Tan, C.H., 2017b. Venomics of *Naja sputatrix*, the Javan spitting cobra: A short neurotoxin-driven venom needing improved antivenom neutralization. *Journal of Proteomics* 157, 18–32.

Tasoulis, T., Isbister, G.K., 2017. A review and database of snake venom proteomes. *Toxins* 9, 290.

Thakur, R., Chattopadhyay, P., Ghosh, S.S., Mukherjee, A.K., 2015. Elucidation of procoagulant mechanism and pathophysiological significance of a new prothrombin activating metalloprotease purified from *Daboia russelii russelii* venom. *Toxicon* 100, 1–12.

Thakur, R., Mukherjee, A., 2015. A brief appraisal on Russell's viper venom (*Daboia russelii russelii*) proteinases. *Snake Venoms*, 1–18.

Trummal, K., Samel, M., Aaspollu, A., Tonismagi, K., Titma, T., Subbi, J., Siigur, J., Siigur, E., 2015. 5'-Nucleotidase from *Vipera lebetina* venom. *Toxicon* 93, 155–163.

Trummal, K., Tonismagi, K., Paalme, V., Jarvekulg, L., Siigur, J., Siigur, E., 2011. Molecular diversity of snake venom nerve growth factors. *Toxicon* 58, 363–368.
Utkin, Y.N., 2013. Three-finger toxins, a deadly weapon of elapid venom--milestones of discovery. *Toxicon* 62, 50–55.
Utkin, Y.N., 2019. Last decade update for three-finger toxins: Newly emerging structures and biological activities. *World Journal of Biological Chemistry* 10, 17–27.
Valerio, A.A., Corradini, A.C., Panunto, P.C., Mello, S.M., Hyslop, S., 2002. Purification and characterization of a phosphodiesterase from *Bothrops alternatus* snake venom. *Journal of Protein Chemistry* 21, 495–503.
Vogel, C.W., Bredehorst, R., Fritzinger, D.C., Grunwald, T., Ziegelmuller, P., Kock, M.A., 1996. Structure and function of cobra venom factor, the complement-activating protein in cobra venom. *Advances in Experimental Medicine and Biology* 391, 97–114.
Vogel, C.W., Fritzinger, D.C., 2010. Cobra venom factor: Structure, function, and humanization for therapeutic complement depletion. *Toxicon* 56, 1198–1222.
Vonk, F.J., Casewell, N.R., Henkel, C.V., Heimberg, A.M., Jansen, H.J., McCleary, R.J., Kerkkamp, H.M., Vos, R.A., Guerreiro, I., Calvete, J.J., Wuster, W., Woods, A.E., Logan, J.M., Harrison, R.A., Castoe, T.A., de Koning, A.P., Pollock, D.D., Yandell, M., Calderon, D., Renjifo, C., Currier, R.B., Salgado, D., Pla, D., Sanz, L., Hyder, A.S., Ribeiro, J.M., Arntzen, J.W., van den Thillart, G.E., Boetzer, M., Pirovano, W., Dirks, R.P., Spaink, H.P., Duboule, D., McGlinn, E., Kini, R.M., Richardson, M.K., 2013. The king cobra genome reveals dynamic gene evolution and adaptation in the snake venom system. *Proceedings of the National Academy of Sciences of the United States of America* 110, 20651–20656.
WHO, 2010. Guidelines for the prevention and clinical management of snake bite in Africa. http://www.afro.who.int/index.php?option=com_docman&task=doc_download&gid=5529
Wijeyewickrema, L.C., Gardiner, E.E., Shen, Y., Berndt, M.C., Andrews, R.K., 2007. Fractionation of snake venom metalloproteinases by metal ion affinity: a purified cobra metalloproteinase, Nk, from *Naja kaouthia* binds Ni2+-agarose. *Toxicon* 50, 1064–1072.
Wong, K.Y., Tan, C.H., Tan, K.Y., Quraishi, N.H., Tan, N.H., 2018. Elucidating the biogeographical variation of the venom of *Naja naja* (spectacled cobra) from Pakistan through a venom-decomplexing proteomic study. *Journal of Proteomics* 175, 156–173.
Wong, K.Y., Tan, K.Y., Tan, N.H., Gnanathasan, C.A., Tan, C.H., 2021. Elucidating the venom diversity in Sri Lankan spectacled cobra (*Naja naja*) through De Novo venom gland transcriptomics, venom proteomics and toxicity neutralization. *Toxins* 13, 558.
Xu, N., Zhao, H.Y., Yin, Y., Shen, S.S., Shan, L.L., Chen, C.X., Zhang, Y.X., Gao, J.F., Ji, X., 2017. Combined venomics, antivenomics and venom gland transcriptome analysis of the monocoled cobra (*Naja kaouthia*) from China. *Journal of Proteomics* 159, 19–31.
Xu, X., Li, B., Zhu, S., Rong, R., 2015. Hypotensive peptides from snake venoms: Structure, function and mechanism. *Current Topics in Medicinal Chemistry* 15, 658–669.
Yamazaki, Y., Morita, T., 2004. Structure and function of snake venom cysteine-rich secretory proteins. *Toxicon* 44, 227–231.
Yamazaki, Y., Morita, T., 2006. Molecular and functional diversity of vascular endothelial growth factors. *Molecular Diversity* 10, 515–527.
Yin, W., Wang, Z.J., Li, Q.Y., Lian, J.M., Zhou, Y., Lu, B.Z., Jin, L.J., Qiu, P.X., Zhang, P., Zhu, W.B., Wen, B., Huang, Y.J., Lin, Z.L., Qiu, B.T., Su, X.W., Yang, H.M., Zhang, G.J., Yan, G.M., Zhou, Q., 2016. Evolutionary trajectories of snake genes and genomes revealed by comparative analyses of five-pacer viper. *Nature Communications* 7, 13107.
Zhang, Y., Wu, J., Yu, G., Chen, Z., Zhou, X., Zhu, S., Li, R., Zhang, Y., Lu, Q., 2011. A novel natriuretic peptide from the cobra venom. *Toxicon* 57, 134–140.
Zhu, W., Smith, J.W., Huang, C.M., 2010. Mass spectrometry-based label-free quantitative proteomics. *Journal of Biomedicine & Biotechnology* 2010, 840518.

6 Envenomation by Asian Cobras

Pathophysiology and Clinical Manifestations

6.1 INTRODUCTION

Due to geographic and species-specific variation in venom composition, Asian Cobra venoms exhibit diverse pharmacological activities under *in vitro* conditions, which can provide important clues for their pathophysiological manifestations in envenomed prey or victims. The following sub-sections highlight some of the *in vitro* pharmacological activities of Asian Cobra venoms and shed light on the diverse clinical presentations of Cobra envenomation in victims. In addition, the differences in the pathophysiology of Cobra envenomation and their severity owing to inter-specific and intra-specific geographical variations have also been correlated with the proteomic composition of the respective Cobra venoms.

6.2 *IN VITRO* PHARMACOLOGICAL PROPERTIES OF ASIAN COBRA VENOMS

The in vitro pharmacological properties of venoms of Indian Spectacled Cobra (*N. naja*) and Indian Monocled Cobra (*N. kaouthia*) were studied in detail in independent laboratories. In a comparative study by Shashidharamurthy and co-workers (2002), Indian Spectacled Cobra (*N. naja*) venom samples from southern, eastern, and western India were evaluated for selected pharmacological properties. The authors found that the venom sample from southern India demonstrates dose-dependent (20–400 μg/ml) *in vitro* anticoagulant activity by prolonging the prothrombin time of citrated human plasma (Shashidharamurthy et al., 2002). This finding was corroborated by a recent study by Attarde et al. (2021) that reported similar anticoagulant properties of southern India *N. naja* venom sample via prolonging the prothrombin time and activated partial thromboplastin time of human platelet-poor plasma (PPP). However, it is to be noted that the authors tested relatively higher doses (100–800 μg/ml) of venom for the assay (Attarde et al., 2021) compared to the earlier study.

The anticoagulant activity of southern India *N. naja* venom can be attributed to the presence of phospholipase A_2 (PLA_2) (~6%), Kunitz-type serine protease inhibitors (KSPI) (~1.4%), and 5'-nucleotidase (5'-NT) (<1%) in this venom as revealed by proteomic analyses (Attarde et al., 2021). Apart from the anticoagulant activity, the southern India *N. naja* venom (10 μg) exhibited indirect hemolytic activity (lysis of

DOI: 10.1201/9781003598121-6

erythrocytes in the presence of phospholipids compared to 100% hemolysis caused by a positive control such as water or a detergent) of 31%. Conversely, the venom sample's direct hemolytic activity (assayed in the absence of phospholipids substrate) was not evaluated. However, a recent study reported that the *N. naja* venom from southern India exhibited appreciable direct hemolysis (~17%) of human erythrocytes (Attarde et al., 2021). The hemolytic activity is primarily caused by PLA_2 enzymes, which are present in moderate amounts (~6%) in this venom sample (Attarde et al., 2021).

Similar to the *N. naja* venom from southern India, the venom sample from western India was also reported to demonstrate dose-dependent (20–400 µg/ml) *in vitro* anticoagulant activity by prolonging the prothrombin time of citrated human plasma (Shashidharamurthy et al., 2002). This finding was corroborated by a recent proteomic study by Chanda et al. (2019), which reported that the western India *N. naja* venom sample can significantly increase the Ca^{2+} clotting time prothrombin time. It activated partial thromboplastin time of goat PPP in a dose-dependent (5–40 µg/ml) manner (Chanda et al., 2019). The anticoagulant activity of western India *N. naja* venom correlates well with the presence of variable amounts of PLA_2 (~7%), KSPI (~1%), and 5'-nucleotidase (5'-NT) (~1.4%) in this venom as revealed by proteomic analysis (Chanda et al., 2019).

Apart from anticoagulant activity, the western India *N. naja* venom (10 µg) also exhibited indirect hemolysis of human erythrocytes, albeit the direct hemolytic activity of the venom sample was not evaluated (Shashidharamurthy et al., 2002). In addition, the western India *N. naja* venom exhibited dose-dependent (5–40 µg/ml) modulation of platelet function; interestingly, lower doses (5 and 10 µg/ml) of the venom-induced aggregation of goat platelet-rich plasma (PRP), while higher doses (20–40 µg/ml) caused de-aggregation of platelets (Chanda et al., 2019). Modulation of platelet function, which is one of the mechanisms to cause hemostatic distress, is primarily induced by isoforms of PLA_2, 5'-NT, L-amino acid oxidase (LAAO), three-finger toxin (3FTx), and snake venom metalloprotease (SVMP) (Dhananjaya et al., 2006; Du and Clemetson, 2002; Dutta et al., 2015; Jagadeesha et al., 2002; Kini and Doley, 2010; Utkin, 2013). The above protein classes were identified in significant proportions in the western India *N. naja* venom (~7% PLA_2, ~1% 5'-NT, ~1% LAAO, ~7% SVMP, ~51% 3FTx) by proteomic analysis (Chanda et al., 2019).

Remarkably, Shashidharamurthy and co-workers (2002) reported that the venom sample from eastern India exhibited dose-dependent (20–200 µg/ml) pro-coagulant activity due to the presence of prothrombin-activating enzymes in this venom sample. However, an in-depth pharmacological characterization of eastern India *N. naja* venom by Dutta et al. (2017) suggested that similar to *N. naja* venom samples from western and southern India, this venom also exhibits dose-dependent (5–40 µg/ml) anticoagulant activity by prolonging the Ca^{2+} clotting time, prothrombin time and activated partial thromboplastin time of goat PPP. In addition, the eastern India *N. naja* venom sample demonstrated dose-dependent (5–40 µg/ml) inhibition of thrombin's fibrinogen clotting activity, further corroborating its anticoagulant property. Moreover, as opposed to the claims of the earlier study (Shashidharamurthy et al., 2002), mass spectrometry analysis of the cation-exchange fractions of the eastern India *N. naja* venom sample did not identify any peptide corresponding to prothrombin activator or thrombin-like protease in the venom (Dutta et al., 2017). On the contrary, the proteome analysis suggested the presence of several anticoagulant protein classes, including phospholipase A_2 (~11% PLA_2), Kunitz-type serine protease inhibitors (~0.5% KSPI), and 5'-nucleotidases (~0.5% 5'-NT). More importantly, the MS/MS analysis suggested the presence of PLA_2 isoforms in eastern India *N. naja* venom that exhibit potent thrombin-inhibitory properties and exert anticoagulant action (Dutta et al., 2017).

Apart from the effect on blood coagulation, the eastern India *N. naja* venom sample also demonstrated modulation of platelet function in goat PRP; it exhibited platelet aggregation at a lower dose (5 μg/ml), while at higher doses (20–40 μg/ml) it could cause de-aggregation of PRP (Dutta et al., 2017). The platelet modulation property of eastern India *N. naja* venom correlates well with the presence of ~11% PLA_2, ~0.5% 5'-NT, ~1% LAAO, ~1% SVMP, and ~64% 3FTx as revealed by proteomic analysis (Dutta et al., 2017).

Das and his co-worker (2013) evaluated some of the pharmacological properties of *N. kaouthia* (Indian monocle Cobra) venom from north-east India (Assam), the major venomous Cobra in this region. They documented that the venom sample demonstrates dose-dependent (0.01–10 μg/ml) anticoagulant activity by prolonging the Ca^{2+} clotting, prothrombin, and activated partial thromboplastin times. The anticoagulant activity of *N. kaouthia* venom from north-east India can be attributed to the presence of significant proportions of PLA_2 (~13%) and 5'-NT (~1%) in this venom as revealed by proteomic analyses (Das et al., 2013; Kakati et al., 2022). In addition to the anticoagulant effect, the *N. kaouthia* venom sample from north-east India exhibited indirect hemolytic activity of ~23% at a 0.5 μg/ml dose in goat erythrocytes. At the same time, it lacked direct hemolytic action up to 5 μg/ml, albeit showed marginal direct hemolysis (~1.4%) at a higher dose of 50 μg/ml (Das et al., 2013).

The anticoagulant property of *N. sagittifera* venom from the Andaman Islands was also recently investigated by Attarde et al. (2021). The authors found that similar to *N. naja* and *N. kaouthia* venom samples from different geographic locations of India, the Andaman Cobra venom (100–800 μg/ml) that contains only 2% PLA_2 enzymes could also prolong the prothrombin time and activated partial thromboplastin time of human PPP and causes blood anticoagulation. Moreover, the venom could also induce indirect hemolysis of 7% in human erythrocytes, although the direct hemolytic activity of the venom sample was not tested (Attarde et al., 2021).

The pharmacological properties, particularly hemostatic deregulation induced by venoms of *Naja* species found in other parts of the Asian continent, including south Asia (Sri Lanka, Bangladesh, Afghanistan, Pakistan, Bhutan, Nepal, Iran), south-east Asia, west Asia, east Asia and central Asia are not well-documented. The majority of the pharmacological properties of Cobra venoms are reported from India. Recently, *N. atra* venom from China containing ~12% PLA_2 enzymes was reported with dose-dependent (10–100 μg/ml) direct hemolysis of ~1–35% in rabbit erythrocytes (Xiao et al., 2022). Nevertheless, it is to be noted that, unlike viperid venoms, the impact of deregulated hemostasis on Cobra venom-induced lethality is minimal, as acute neurotoxicity is the primary pathophysiological manifestation of Cobra envenoming.

Among the reported pharmacological properties, significant variations can be observed among different species of Asian Cobras, mainly due to inter-specific variations in the venom composition. However, the observed variation in some pharmacological properties among the Cobra venoms can also be due to different assay conditions. For example, the incubation time for direct hemolysis in the assay protocols was significantly different for *N. naja*, *N. kaouthia*, *N. sagittifera*, and *N. atra*. While *N. kaouthia* and *N. atra* venoms were incubated for 60 min with goat and rabbit erythrocytes, respectively, the same for *N. naja* and *N. sagittifera* venoms was 24 h with human erythrocytes. It is well known that the membrane phospholipid composition of erythrocytes from different mammalian species may vary. Therefore, apart from the variable distribution of PLA_2 enzymes (responsible for hemolysis), a difference in incubation time and the source of erythrocytes can also contribute to the observed variation in direct hemolytic properties of the above Asian Cobra venoms. Therefore, it will be interesting to note the results when the assays are performed under uniform experimental conditions, preferably in rodent erythrocytes, since they are the natural prey of the Cobras.

In another interesting proteomic study by Tan and his co-workers (2019), the authors investigated the cytotoxic effect of venoms of *N. philippinensis* (Philippines), *N. sumatrana* (Malaysia), *N. kaouthia* (Malaysia), and *N. atra* (Taiwan) on murine L (fibroblast) cells and human 184B5 (epithelial) cells using an MTT (3-(4,5-dimethylthiazol-2-yl)-2,5-diphenyltetrazolium bromide)-based assay. All the tested venoms exhibited cytotoxicity to both the cell lines, albeit with variable potency and specificity. In the murine cell line, the cytotoxic potency of *N. philippinensis* venom (IC_{50} = 5.6 μg/ml) was the highest, followed by venoms of *N. atra* (IC_{50} = 6.9 μg/ml), *N. sumatrana* (IC_{50} = 10.8 μg/ml) and *N. kaouthia* (IC_{50} = 20.1 μg/ml). However, in the human epithelial cell line, the *N. sumatrana* venom (IC_{50} = 14.3 μg/ml) was found to be the most cytotoxic, followed by venoms of *N. kaouthia* (IC_{50} = 15.3 μg/ml), *N. atra* (IC_{50} = 18.0 μg/ml) and *N. philippinensis* (IC_{50} = 18.7 μg/ml) (Tan et al., 2019). It is evident from these results that with lower IC_{50} values, the murine cells are more susceptible to the above *Naja* venoms, perhaps because they are the natural prey for the Cobras.

Furthermore, all the tested Cobra venoms contain significant proportions of cytotoxic proteins, including cytotoxins (~20–60%), PLA_2 (~12–35%), SVMP (~2–5%) and LAAO (~1%) that corroborates their cytolytic properties (Huang et al., 2015; Tan et al., 2022; Tan et al., 2019; Tan et al., 2015). Furthermore, the results highlight inter-specific differences in the cytotoxic properties of venom in different Cobra species from Asian countries.

6.3 VARIOUS FACTORS PRESENT IN THE VENOM OF ASIAN COBRAS AFFECT THEIR LETHALITY

Asian Cobras are front-fanged snake species that have evolved with a specialized venom delivery system to inject vast amounts of venom (150 to 250 mg). Specific Cobra venom toxins employ local pathophysiological effects in the adjoining tissues at the bite site upon envenomation. In contrast, others are disseminated via the blood, vascular, and lymphatic systems to various organs, resulting in systemic clinical manifestations (Gutierrez et al., 2017). Nevertheless, the presentation of envenomation syndromes due to Asian Cobras can be highly variable, and it greatly depends on the inflicting Cobra species, geographic, seasonal, and ontogenic variation in venom composition, and the amount of venom injected during the envenoming.

Several methods have been developed to evaluate the toxicity of snake venoms; however, the most widely accepted method is determining the median lethal dose (LD_{50} value) in a live mouse model. The LD_{50} values of diverse Asian Cobras are listed in Table 6.1. It is to be noted that the reported LD_{50} value of the venom of a particular species may vary across independent studies. These differences primarily arise due to geographic, seasonal, and ontogenic venom variation and other factors, including methods of venom collection, storage, reconstitution, mice or rat strain, and venom injection route during LD_{50} determination. Among the various major toxins in Asian Cobra venoms, the lethality of αNTXs (LD_{50} ~0.05–0.2 μg/g) is the highest (Palasuberniam et al., 2021; Tan et al., 2019). Therefore, the lethal potency of Asian Cobra venoms (LD_{50} values) exhibits a fair correlation with the proportions of αNTXs present in the respective venom proteomes (Figure 6.1). For most Asian Cobra venom samples, it was observed that when the relative abundance of αNTXs is ≤20%, the LD_{50} value for that venom sample is ≥0.5 μg/g. On the contrary, if the relative abundance of αNTXs is ≥20%, the LD_{50} value is ≤0.5 μg/g.

However, *N. naja* venom from southern India and *N. sumatrana* venom samples from Sumatra and Thailand stand out as exceptions. The *N. naja* venom from southern India is predominated by αNTXs (~51%); however, the relative lethality of this venom sample is relatively lower (LD_{50} ~0.8 μg/g) compared to *N. naja* venoms from other geographical locations of the Indian sub-continent (Table 6.1, Figure 6.1). Conversely, the venom samples of *N. sumatrana* exhibited profound lethality (LD_{50} ~0.4 μg/g) despite containing relatively lower amounts of αNTXs (~10% to 16%). This data suggests that apart from the relative proportion of a particular primary toxin class in a venom sample, the lethal potency of individual toxin isoforms also contributes to the overall toxicity of the venom.

Another critical factor that directly impacts the severity of clinical manifestation is the amount of venom injected by the offending snake during envenomation. It is reasonable to believe that a venomous snake does not consider human beings as prey but as a threat, and snake envenoming is self-defense against this imposing threat. The venom production in the venom glands is metabolically costly; thus, to conserve venom for their prey, the offended individuals may occasionally deliver warning bites devoid of venom (Morgenstern and King, 2013). These encounters are often termed as ‘dry bites’, and the strategy adopted by snakes for the optimum usage of venom is known as ‘venom metering’ (Hayes et al., 2002). About 20% to 30% of Cobra bites are estimated to result in dry bites (Pucca et al., 2020). It is also noteworthy that the phenomenon of ‘venom metering’ is more frequently observed in adult Cobra snakes, which can control the venom gland musculature (mainly accessory venom gland) to avoid excessive venom expulsion. On the contrary, the juveniles and neonates tend to empty their glands during envenoming (Pucca et al., 2020).

In addition to deliberate control of venom expulsion, the amount of venom present in the glands during the envenomation incident can also determine the severity of clinical presentations. A few studies have shown that venom replenishment may take

TABLE 6.1
Summary of the LD_{50} Values (μg/g) of Asian Cobra Venoms Determined in Diverse Animal Models Under In Vivo Conditions

Cobra Species	Venom Source	LD_{50} Value (i.v.) (μg/g)	Animal Model	Reference
N. atra	Eastern Taiwan	0.67	ICR mice	Huang et al. (2015)
	Western Taiwan	0.33	ICR mice	Huang et al. (2015)
	Zhejiang, China	0.68	ICR mice	Shan et al. (2016)
N. kaouthia	Thailand	0.24	CD-1 mice	Laustsen et al. (2015)
	Malaysia	0.90	ICR mice	Tan et al. (2015)
	Thailand	0.18	ICR mice	Tan et al. (2015)
	Vietnam	0.90	ICR mice	Tan et al. (2015)
	Thailand	0.34	C57BL/6Narl strain	Liu et al. (2017)
	Guangxi, southern China	0.79*	ICR mice	Xu et al. (2017)
	Burdwan, eastern India	0.70	BALB/c mice	Mukherjee and Maity (2002)
	Assam, North-east India	0.15*	Swiss albino mice	Das et al. (2013)
N. naja	Sindh, Pakistan	0.22	ICR mice	Wong et al. (2018)
	Burdwan, eastern India	0.40	BALB/c mice	Mukherjee and Maity (2002)
	West Bengal, eastern India	0.70*	Swiss Wistar mice	Shashidharamurthy et al. (2002)
	Maharashtra, western India	1.20*	Swiss Wistar mice	Shashidharamurthy et al. (2002)
	Irula, southern India	2.00*	Swiss Wistar mice	Shashidharamurthy et al. (2002)
	Karnataka, southern India	0.84	CD-1 mice	Attarde et al. (2021)
	Colombo, Sri Lanka	0.75	ICR mice	Wong et al. (2021)
N. philippinensis	Southern Philippines	0.18	ICR mice	Tan et al. (2019)
N. sagittifera	Port Blair, Andaman Island	0.48	CD-1 mice	Attarde et al. (2021)
N. samarensis	Southern Philippines	0.20	ICR mice	Palasuberniam et al. (2021)
N. siamensis	Thailand	0.46	C57BL/6Narl strain	Liu et al. (2017)
N. sputatrix	Java Island	0.90	ICR mice	Tan et al. (2017)
N. sumatrana	Negeri Sembilan, Malaysia	0.50	ICR mice	Tan et al. (2022)
	Penang, Malaysia	0.96	ICR mice	Tan et al. (2022)
	Southern Thailand	0.41	ICR mice	Tan et al. (2022)
	Sumatra	0.39	ICR mice	Tan et al. (2022)

* Indicates intra-peritoneal (i.p.) LD_{50} values.

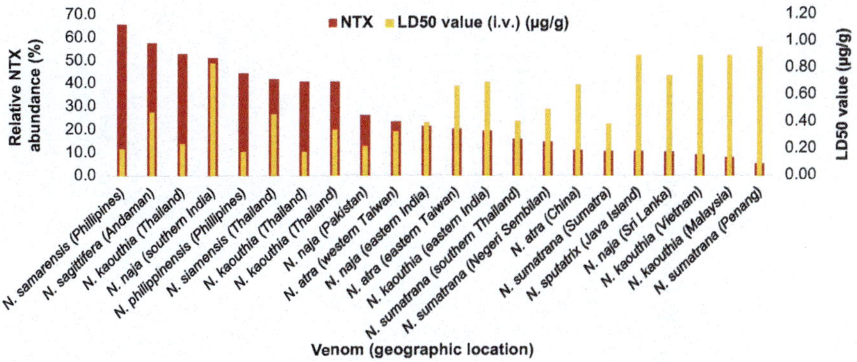

FIGURE 6.1 A bar diagram depicting the correlation between proportions of neurotoxins (αNTXs) present and reported LD_{50} values (μg/g) in diverse Asian Cobra venoms.

up to 14 days once the venom glands are emptied, and any envenomation during this recovery period will naturally deliver lower amounts of venom into the victims (McCue, 2006; Yin et al., 2020). Interestingly, a recent study by Harris and co-workers (2020) showed variable composition in venoms extracted from the left and right glands of certain specimens of Indochinese Spitting Cobra (*Naja siamensis*). The authors hypothesized that the differences in the venom composition may be due to differential rates of toxin replenishment in each of the glands post-ejection from the glands during envenoming (Harris et al., 2020).

6.4 PATHOPHYSIOLOGY OF ASIAN COBRA ENVENOMATION: CLINICAL PRESENTATION AND RESPONSIBLE TOXINS

Similar to most elapids, envenomation by Asian Cobras is primarily characterized by neurotoxicity and local tissue damage of variable extents depending upon the qualitative and quantitative distribution of neurotoxins and cytotoxic components in the venoms (Bawaskar and Bawaskar, 2004; Gutierrez et al., 2017; Kakati et al., 2023). The following sub-sections elaborate on the sequence of these predominant clinical presentations of Cobra envenomation and try to correlate these pathophysiologies with the constituent Cobra venom proteins. The major clinical manifestations (neurotoxicity and cytotoxicity) of Asian Cobra envenomation and the responsible toxins are also summarized in Table 6.2.

6.4.1 Neurotoxicity

Neurotoxicity is characterized by systemic neuromuscular paralysis and is a hallmark clinical presentation in Asian Cobra envenomation (Gutierrez et al., 2017). Most envenomed patients demonstrate signs of progressive paralysis, which commences with the manifestation of bilateral ptosis (drooping of the upper eyelids) within an hour of envenomation. This symptom is usually followed by progressive external ophthalmoplegia (weakness of the eye muscles), disturbed vision, limited

TABLE 6.2
The Major Clinical Manifestations of Asian Cobra Envenomation and the Relative Distribution of the Toxins Responsible for These Pathophysiologies

Pathophysiology of Envenomation	Species	Study Location	Total Number of Cases	Frequency of the Clinical Manifestation	Responsible Toxins and Their Relative Abundance (% of Venom Protein)	References
					NTX	
Neurotoxicity	*N. atra*	Hong Kong, Taiwan	225	5–20%	11–20%	Mao et al. (2018), Wong et al. (2010), Hung et al. (2003)
	N. kaouthia	India, Bangladesh (Indian sub-continent)	89	80–85%	15–20%	Faiz et al. (2017), Kakati et al. (2023)
	N. kaouthia	Thailand (South-east Asia)	565	30–55%	40–50%	Buranasin (1993), Wongtongkam et al. (2005), Tansuwannarat et al. (2023)
	N. mandalayensis	Myanmar	15	33%	Not available	Sai Sein Lin et al. (2020)
	N. naja	India, Sri Lanka	69	40–60%	10–20%	Kularatne et al. (2009), Bawaskar and Bawaskar (2004), Bawaskar et al. (2007)
	N. philippinensis	Philippines	39	97%	45%	Watt et al. (1988)
	N. siamensis	Vietnam	17	10%	42%	Quyen (2004)

					CTX	PLA2	SVMP	LAAO	
Cytotoxicity, local swelling, tissue necrosis, and pain at the bite site	*N. atra*	Hong Kong, Taiwan	225	85–95%	53–65%	12–17%	1.5–2.7%	0.2%	Mao et al. (2018), Wong et al. (2010), Hung et al. (2003)
	N. kaouthia	Indian subcontinent (India, Bangladesh)	89	75–100%	20–44%	12–13%	4.5–5.3%	0.1–0.8%	Faiz et al. (2017), Kakati et al. (2023)
	N. kaouthia	Southeast Asia (Thailand)	565	70–100%	20–27%	13–17%	2.6–4.4%	0.3–0.4%	Buranasin (1993), Wongtongkam et al. (2005), Tansuwannarat et al. (2023)
	N. mandalayensis	Myanmar	15	40%	Not available				Sai Sein Lin et al. (2020)
	N. naja	India, Sri Lanka	69	95–100%	25–70%	10–27%	4.0–6.6%	0.5–1.9%	Kularatne et al. (2009), Bawaskar and Bawaskar (2004), Bawaskar et al. (2007)
	N. philippinensis	Philippines	39	8%	21%	23%	3.9%	0.4%	Watt et al. (1988)
	N. siamensis	Vietnam	17	88%	34%	15%	1.59%	0.5%	Quyen (2004)

mouth movement and opening, slurred speech, and frothy saliva due to deficiency of the facial muscles. Gradually, the paralysis descends to the bulbar muscles (present in the mouth and throat for assisting speech and swallowing), causing difficulty in swallowing. After that, the patients experience paralysis of the neck flexor muscles, resulting in a 'broken neck' sign. Finally, the respiratory muscles and diaphragm of the patients are paralyzed, inducing a state of hypoventilation, which culminates in respiratory failure if not treated promptly (Bawaskar et al., 2017; Gutierrez et al., 2017; Gutierrez et al., 2006; Hamza et al., 2021).

The degree of neuromuscular paralysis in Asian Cobra envenoming can vary from mild to severe depending upon the amount of venom reaching the target sites as well as the composition of venom of the inflicting species. Primarily, the αNTXs present in Asian Cobra venoms are responsible for inducing neurotoxic symptoms in envenomed patients (Table 6.2). This class of 3FTxs can cause blockade of neurotransmissions at the neuromuscular junctions by binding post-synaptically to the muscle nAChRs and neuronal α7 nAChRs. This high-affinity association of αNTXs with nAChRs subsequently blocks the binding of endogenous acetylcholine neurotransmitters to their receptors, causing flaccid paralysis (Nys et al., 2022; Ranawaka et al., 2013).

6.4.2 Cytotoxicity and Local Effects

Apart from systemic neurological manifestations, local tissue damage is another major clinical presentation in Cobra envenoming that is challenging to mitigate (Bawaskar and Bawaskar, 2004; Quyen, 2004). The local effects of Asian Cobra envenomation are frequently characterized by acute radiating pain and progressive swelling (edema) at the bite site, blistering, local tissue ischemia, chronic non-healing ulcer formation, and extensive soft tissue necrosis that can cause lifelong morbidities in victims (Bawaskar et al., 2017; Gutierrez et al., 2017; Gutierrez et al., 2006; Hamza et al., 2021). The presentation of these local signs is primarily attributed to CTXs, a sub-class of the 3FTx family with considerable abundance in Asian Cobra venoms (Table 6.2).

However, it has been observed that while the proportions of αNTXs dictate the severity of neurotoxic symptoms in most Cobra-envenomed patients, the relative abundance of CTXs does not share a discrete correlation with the extent of local tissue damage. This data indicates the existence of CTX isoforms with variable cytotoxic potency and the involvement of other Cobra venom toxins that can induce these local effects alone or in association with CTXs. Once envenomation occurs at a particular site, most CTXs concentrate at the bite site owing to their high affinity binding to specific cell membrane phospholipid micro-domains. After that, they deploy diverse mechanisms that progressively result in cellular damage near the bite site (Gutierrez et al., 2017).

Cobra venom CTXs can penetrate the cell membrane and induce disruption of membrane integrity, thereby leading to cytolysis (Dubovskii et al., 2014; Konshina et al., 2012). In another mechanism, post-cellular penetration, CTXs are localized in the lysosomes and mitochondria, which induces perturbation of the organellar membranes, leading to the activation of necrotic and apoptotic cell death pathways, respectively (Zhang et al., 2019). Notably, some CTX isoforms can directly activate pro-apoptotic proteins (Bax and Bak) as well as caspases (3 and 9) that subsequently induce apoptotic cell death (Chen et al., 2008; Chen et al., 2009; Yang et al., 2007). These mechanisms ultimately result in local tissue necrosis (Kalita et al., 2022). Apart from the CTXs, certain Cobra venom PLA_2s, SVMPs, and LAAOs that occur in variable amounts in Asian Cobra venoms (Table 6.2) can also contribute to local tissue damage by diverse mechanisms (Gutierrez et al., 2006; Lomonte et al., 2003; Tan and Tan, 2015). Interestingly, although most Cobra venom PLA_2s are non-toxic in isolation, they can form cognate complexes with CTXs that can potentiate the action of the latter and aggravate local tissue necrosis (Dutta et al., 2019; Mukherjee, 2010). SVMPs primarily target specific proteins of the basement membrane of capillaries at the bite site for degradation, thereby augmenting tissue damage (Escalante et al., 2011; Gutierrez and Rucavado, 2000). Further, LAAOs are known for their ability to induce cellular damage by generating reactive oxygen species (Mukherjee et al., 2015; Mukherjee et al., 2018; Tan and Tan, 2015).

Cobra venom PLA_2s are also responsible for the progressive swelling at the bite site. By their catalytic property, these enzymatic Cobra venom toxins can hydrolyze membrane phospholipids, leading to the generation of pro-inflammatory lipid mediators, including thromboxanes, prostacyclins, leukotrienes, and prostaglandins that can subsequently induce edema at the bite site (Estevao-Costa et al., 2016; Gutiérrez et al., 2016; Teixeira et al., 2009). Moreover, enzymatic degradation of membrane phospholipids by Cobra venom PLA_2s generates pain-inflicting compounds such as prostaglandins, biogenic amines, and bradykinin (Zhang et al., 2017). In addition, PLA_2s can augment the CTX-mediated activation of sensory neurons, thereby inflicting radiating pain at the bite site of the envenomed victims (Kazandjian et al., 2021).

6.4.3 Cardiotoxicity

Cardiac perturbations and myocardial injury have been observed in Cobra-envenomed cases from the Asian continent (Cher et al., 2005; Wongtongkam et al., 2005). These detrimental cardiac effects ultimately lead to the development of myocardial ischemia, dysrhythmia, and even cardiac arrest in some victims. Cardiotoxic effects of Cobra venoms have been associated with CTXs (Gasanov et al., 2015; Harvey et al., 1983; Raynor et al., 1991), although the indirect involvement of other venom toxins, including PLA_2s and NTXs cannot be ruled out (Averin and Utkin, 2021; Wu, 1998).

Apart from localizing at the bite site and inducing local tissue damage, CTXs can also disseminate from the bite site and reach the heart muscles of the victims via the blood-vascular system. These CTXs can exert cytotoxic effects on the cardiomyocytes, eventually leading to cardiotoxicity. Therefore, CTXs are also often called cardiotoxins (CdTxs) (Slotta and Vick, 1969). In mechanism, the CdTxs either specifically activate the calcium channels in the cardiomyocyte membrane or form non-selective pores in the membrane to alter the intracellular calcium concentration of the cardiomyocytes. This event subsequently leads to heart muscle contraction alterations that culminate in cardiac arrests in Cobra-envenomed victims (Averin et al., 2022; Kwan et al., 2002; Wang and Wu, 2005).

In addition to CdTxs, Cobra venom PLA_2s possessing hypotensive properties can cause an increase in capillary membrane permeability, which has cardiotoxic consequences (Castro-Amorim et al., 2023; Peterfi et al., 2019). In addition, NTXs cause a state of hypoxia in different organs, including the heart, by inducing respiratory distress that can also contribute to cardiotoxicity (Zhao et al., 2017).

6.4.4 Ophthalmia

Venom ophthalmia is a rare clinical condition characterized by exposure of a victim's eye to the venom of a spitting Cobra species (ocular envenoming). The venom-spitting behavior is prevalent in seven Asian Cobra species, including *N. atra, N. sumatrana, N. siamensis, N. sputatrix, N. philippinensis, N. mandalayensis* and *N. samarensis*. This unique spitting behavior of specific Asian Cobra species is considered a projectile defense mechanism, and humans usually get accidentally affected. It has been reported that certain Cobras can spit venom in the form of jet spray up to a distance of 2 to 3 meters (Kazandjian et al., 2021; Slowinski and Wüster, 2000). During such events, if the venom comes in contact with the human eyes, it induces acute and intense pain with a burning sensation. This event can be followed by swelling of the eyelids, congestion of the conjunctiva, corneal erosion, ocular discharge, and blurring of vision (Chang et al., 2020; Chu et al., 2010). Notably, suppose the condition is not attended to in due time, the patient may suffer from temporary or permanent blindness due to corneal opacity—a detailed chronology of the clinical features post-ocular envenoming is depicted in Figure 6.2.

Although epidemiological studies on the clinical manifestation of envenomation by Asian Cobra species are well-documented, the same is not valid for venom ophthalmia. A few rudimentary studies have suggested venom spitting and subsequent ophthalmia in some of the Asian Cobra species, including *N. mandalayensis* (Burma), *N. sumatrana* (Sumatra, Malaysia, Thailand, and Bormeo), *N. siamensis* (Burma, Cambodia, Thailand and Lao PDR), *N. sputatrix* (Java), *N. philippinensis* (Philippines), *N. samarensis* (Philippines) and *N. atra* (China); however, these studies lack detailed quantitative data (Chu et al., 2010).

A recent epidemiological study in Taiwan from 1990 to 2016 found 39 patients suffering from toxic eye disease caused by *N. atra* (Tsai et al., 2020). The retrospective data suggested that ophthalmia was primarily reported in males (95%) of the age group 18 to 59 years old (90%), and the majority of these cases (82%) involved venom spitting into a single eye. As expected, these encounters primarily happened during an attempt made by the victim to catch the snake (51%). Further, these cases were reported mainly during the spring and summer (92%). Clinically, 90% of the victims developed acute ocular symptoms, including redness of the affected eye (85%), ocular pain (90%), corneal injury (59%), and conjunctivitis (67%).

Another recent epidemiological study retrospectively analyzed 506 cases of venom ophthalmia caused by *N. siamensis* and *N. sumatrana* from 2018 to 2021 in Thailand (Tansuwannarat et al., 2023). The data indicated that males (79%) were primarily affected, and almost all the cases (503/506) involved venom spitting into a single eye. Furthermore, 94% of the victims reported irritation of the affected eye(s),

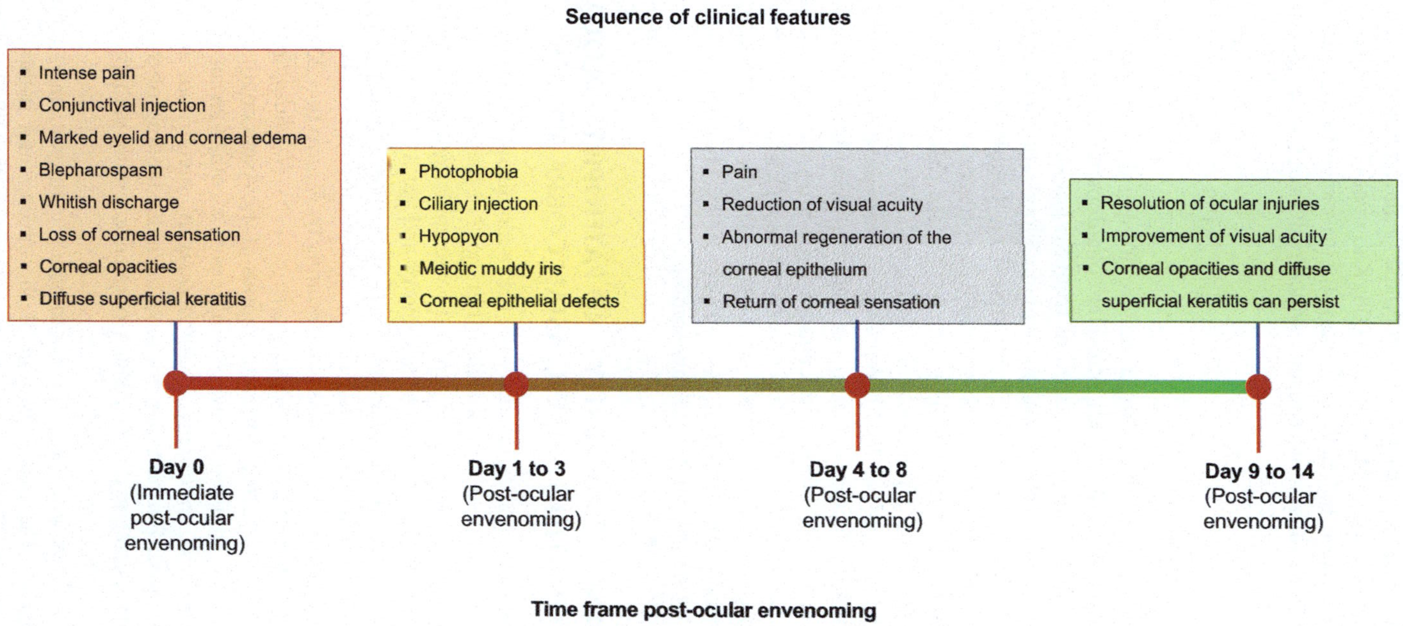

FIGURE 6.2 A schematic diagram of the sequence of clinical presentations during cobra venom ophthalmia.

and 10% suffered a decrease in visual acuity. These clinical presentations of ocular envenomation have been attributed to synergistic or independent actions of CTXs, PLA_2s, SVMPs, and LAAOs (Kao et al., 2008; Kao et al., 2009; Wang et al., 2006). As discussed in Section 6.3.2, these venom toxins can cause tissue damage and induce pain at the site of venom exposure, which results in ocular damage in venom ophthalmia.

6.4.5 Coagulopathy

While coagulopathy is a distinct clinical feature of viperid envenomation (Gutierrez et al., 2006; Mukherjee et al., 2000), transient coagulopathy has only been reported in a few cases of Cobra-envenomed victims from Sri Lanka (Kularatne et al., 2009). Coagulotoxic manifestations include incoagulable blood, spontaneous bleeding, and systemic capillary leak syndrome. These hemostatic disturbances are the result of the concerted action of SVMPs, SVSPs, and PLA_2s present in snake venoms (Castro et al., 2004; Dutta et al., 2015; Mukherjee, 2008, 2014; Mukherjee and Mackessy, 2013; Saikia et al., 2012; Suntravat et al., 2010; Takeya et al., 1992; Thakur et al., 2015; Thakur and Mukherjee, 2015). While the proteases catalytically activate several blood coagulation factors that result in their depletion from the blood vascular system, PLA_2s can cause blood anticoagulation by enzymatic and non-enzymatic mechanisms. Nevertheless, it is to be noted that except PLA_2s, the relative abundances of SVMPs (~1 to 10%) and SVSPs (~0.1 to 1%) in Asian Cobra venoms are pretty marginal compared to that of the 3FTxs (>50%). Therefore, coagulopathy is rarely expected to manifest in Cobra envenoming.

6.5 DIFFERENCES IN SEVERITY AND PATHOPHYSIOLOGY OF ENVENOMATION AMONG ASIAN COBRA SPECIES

Although neurotoxicity and cytotoxicity are the predominant clinical manifestations of Asian Cobra envenoming, remarkable differences in the severity of these pathophysiologies have been documented across diverse epidemiological studies from the Asian continent. However, to prevent ambiguity and get a clear picture of the disparities, we have restricted our discussions to only those epidemiological studies that have reported confirmed cases of Cobra envenoming. It is to be noted that we could retrieve 1018 cases of confirmed envenoming by *N. atra* (n = 225), *N. kaouthia* (n = 654), *N. naja* (n = 68), *N. mandalayensis* (n = 15), *N. philippinensis* (n = 39), and *N. siamensis* (n = 17) from different regions of the Asian continent (Table 6.2). Unfortunately, many confirmed envenoming cases of other Asian Cobra species, including *N. oxiana*, *N. sagittifera*, *N. samarensis*, *N. sumatrana*, and *N. sputatrix* are yet to be reported. Furthermore, for ease of understanding of the inter-specific disparities, we have categorized the Asian Cobra envenoming pathophysiologies into three groups –

(i) predominantly neurotoxic, (ii) predominantly cytotoxic, and (iii) both neurotoxic and cytotoxic. The categorization of the envenoming pathophysiologies is based on the frequency of the respective clinical presentation (Table 6.2).

6.5.1 The Venom of the Asian Cobra, Which Presents Mainly Neurotoxic Symptoms

Although most Cobra envenomed victims manifest at least some neurological disorders, neuro-paralysis is the predominant clinical presentation in *N. philippinensis* envenomation. Nearly 97% of the reported envenomed victims (n = 39) developed neurotoxicity with minimal local tissue damage, observed in only 8% of the patients (Table 6.2). Further, 19 patients suffered severe neurotoxicity with rapid onset of respiratory paralysis, which could be recorded as early as 30 minutes post-envenomation in 3 cases (Watt et al., 1988).

The manifestation of rapid and severe neurotoxicity in *N. philippinensis* envenomed cases can be well-explained based on the proteome composition of this Asian Cobra species, which is predominated by potent short-chain αNTX (45%) (Table 6.2). The mild local tissue damage observed in a few envenomed cases can be attributed to the occurrence of significant proportions of CTX (21%) and PLA_2 (23%) in its venom (Table 6.2). It is also interesting to note that despite being a spitting species, *N. philippinensis* venom contains relatively lower amounts of CTXs compared to some other Asian spitting Cobras (~40–50%) and, as a result, does not inflict severe cytotoxicity. Moreover, the potency of CTXs isolated from *N. philippinensis* venom also seems relatively lower. A study by Tan et al. (2019) showed that CTXs isolated from *N. philippinensis* venom exhibit relatively lower levels of cytotoxicity (IC_{50} = 18.7 µg/ml) against a human mammary tissue cell line (184B5) compared to purified CTXs of other Asian Cobras including, *N. atra* (IC_{50} = 18.0 µg/ml), *N. kaouthia* (IC_{50} = 15.3 µg/ml) and *N. sumatrana* (IC_{50} = 14.3 µg/ml) (Tan et al., 2019). Nevertheless, the evolutionary significance of producing such a potent neurotoxic venom by a spitting Cobra species is yet to be unraveled.

6.5.2 Asian Cobra Envenomation That Predominantly Presents Cytotoxicity and Local Tissue Damage

Envenomation by *N. atra* and *N. siamensis* primarily manifests extensive local tissue damage with infrequent neurological disorders (Hung et al., 2003; Mao et al., 2018; Quyen, 2004; Wong et al., 2010). In the case of the former Asian Cobra species, cytotoxicity, local swelling, and extensive local tissue necrosis were reported in 85 to 95% of the envenomed victims (n = 225). However, only 5 to 20% of the patients presented with mild to moderate neurotoxic symptoms. Similarly, in *N. siamensis* envenomed cases (n = 17), cytotoxicity was reported in 88% of the patients, while only 10% of the victims exhibited mild neurotoxic symptoms (Table 6.2).

The frequent manifestation of severe local tissue necrosis in *N. atra* envenomed victims is consistent with the venom proteome composition; the venom of this spitting Cobra species is predominated by CTXs (53–65%) and PLA_2s (12–17%). On the contrary, *N. atra* venom contains relatively lower amounts of NTXs (10–20%) and is responsible for the infrequent and moderate neurotoxic symptoms in envenomed victims (Table 6.2). However, similar to the *N. philippinensis* venom proteome, the venom of *N. siamensis* contains significant proportions of αNTXs (42%), yet neurotoxicity is rarely observed in bite victims. This data suggests that the αNTX isoforms of *N. siamensis* venom are less potent than those of the *N. philippinensis* venom. However, this still needs to be experimentally validated.

Furthermore, the *N. siamensis* venom contains moderate amounts of CTXs (34%) and PLA_2s (15%), which are responsible for the observed severe local clinical manifestations (Table 6.2). Following the general assumption, the prominent cytotoxic effects of venoms from these two Asian Cobra species (*N. atra* and *N. siamensis*) are expected due to their spitting behavior. It is also interesting to note that although *N. siamensis* envenomed victims rarely exhibit neurotoxic symptoms, the *N. siamensis* venom used for the proteome analysis could induce both neuro-paralysis and severe tissue necrosis in the experimental mouse model (Liu et al., 2017). These distinct clinical symptoms in humans and mice (natural prey) hint at the prey-specific evolution of constituent Cobra venom toxins.

6.5.3 Asian Cobra Venom That Exerts Both Neurotoxic and Cytotoxic Pathophysiology

While envenomation by the above-described Asian Cobra species predominantly manifests either neurotoxic or cytotoxic symptoms, bites by *N. kaouthia*, *N. naja*, and *N. mandalayensis* are complex and are characterized by both local and systemic pathophysiologies (Bawaskar and Bawaskar, 2004; Bawaskar et al., 2008; Buranasin, 1993; Faiz et al., 2017; Kakati et al., 2023; Kularatne et al., 2009; Sai Sein Lin et al., 2020; Wongtongkam et al., 2005).

In *N. naja* envenomation cases (n = 69) reported from India and Sri Lanka, the frequencies of neurotoxicity and cytotoxic presentations were recorded at 40–60% and 95–100%, respectively (Table 6.2). *N. naja* venoms contain significant amounts of CTXs (25–75%), PLA_2s (10–27%), and SVMPs (4–6%) that are responsible for the extensive local tissue damage observed in envenomed victims. In addition, the venoms contain moderate amounts of αNTXs (10–20%) that cause neuro-paralysis in envenomed victims (Attarde et al., 2021; Chanda et al., 2019; Chanda and Mukherjee, 2020; Chanda et al., 2018; Dutta et al., 2017; Sintiprungrat et al., 2016; Wong et al., 2018; Wong et al., 2021). Similarly, in *N. mandalayensis* envenomed victims (n = 15), neurotoxicity was observed in 33% of the patients, while 40% of the cases presented with local tissue damage. Unfortunately, the classification of toxin families of the *N. mandalayensis* venom proteome was done based on 'UniProt molecular function' and 'mechanism of action' identifiers (Beraldo Neto et al., 2021) rather than on snake venom protein families and therefore, the observed clinical manifestations could not be unambiguously correlated.

Interestingly, although *N. kaouthia* envenomed cases developed both neurotoxicity and cytotoxicity, there seems to be a striking difference in the severity and frequency of neurological disorders among victims from the Indian sub-continent and Southeast Asia. While the frequency of neurotoxic manifestations in envenomed cases (n = 89) from India and Bangladesh (Indian sub-continent) was found to be 80–85%, the same was significantly lower (30–55%) in *N. kaouthia* envenomed cases (n = 565) from Thailand (Southeast Asia). However, the relative distribution of αNTXs in *N. kaouthia* venom from the Indian sub-continent and Southeast Asia is significantly different, and its correlation to the frequency of manifestation of neurotoxic symptoms is quite perplexing. While the proportions of αNTXs in *N. kaouthia* venom from the Indian sub-continent (15–20%) is relatively lower, the frequency of neurotoxic manifestations is significantly higher (80–85%).

On the contrary, despite having a higher relative abundance of αNTXs (40–50%) in its venom, the frequency of neurological manifestations in *N. kaouthia* envenomed victims from Southeast Asia is significantly lower (30–55%) (Table 6.2). These clinical data restore the opinion that the severity of a particular clinical manifestation depends not exclusively on the relative abundance but also on the potency of the toxins responsible. Although this has been experimentally verified for CTXs of some Asian Cobra venoms (Tan et al., 2019), the same is yet to be explored for other Cobra venom toxins, including αNTXs. Furthermore, the evolutionary significance of such tremendous disparities in relative abundances and potencies (if any) of *N. kaouthia* venom toxins needs the close attention of the toxinologists.

Nevertheless, the frequent presentation of extensive local tissue damage in *N. kaouthia* envenomation was consistent among cases reported from the Indian sub-continent (75–100%) as well as Southeast Asia (70–100%) (Table 6.2). The CTXs, PLA_2s, and SVMPs constitute significant proportions in *N. kaouthia* venom from the Indian sub-continent (20–44%, ~13%, and ~5%, respectively) as well as Southeast Asia (20–27%, 13–17%, and ~4%, respectively) (Table 6.2). Therefore, these cytotoxic Cobra venom toxins are expected to cause persistent local tissue necrosis experienced by *N. kaouthia* envenomed victims from both locations.

6.6 CONCLUSION

In vitro pharmacological properties of some Asian Cobra venoms, particularly effects on blood homeostasis, have been reported. Notably, some interspecific differences exist in pharmacological properties, primarily due to variations in venom composition among Asian Cobra species. Nevertheless, the influence of variable assay conditions on the observed differences cannot be ruled out. Furthermore, it is also important to note that Cobra venom is primarily associated with neurotoxicity and cytotoxicity, and hemostatic deregulation may be a secondary clinical manifestation. The lethality of Cobra venoms in terms of median lethal dose (LD_{50} value) assayed in mouse models has also been investigated across several Asian species, and the findings suggest a strong positive correlation between LD_{50} value and relative distribution of αNTXs. Epidemiological studies on Cobra envenomation across Asian countries suggest highly variable clinical manifestations. Some Cobra species primarily inflict neurotoxic symptoms without causing significant local effects (*N. philippinensis*), some are predominantly cytotoxic (*N. atra* and *N. siamensis*), and a few others cause both systemic and local effects (*N. naja*, *N. kaouthia*, and *N. mandalayensis*). This dramatic variation in clinical presentation among Asian Cobras is dictated by the qualitative and quantitative distribution of major toxins, including CTXs, αNTXs, and PLA_2 in their venom proteomes.

REFERENCES

Attarde, S., Khochare, S., Iyer, A., Dam, P., Martin, G., Sunagar, K., 2021. Venomics of the enigmatic Andaman Cobra (*Naja sagittifera*) and the preclinical failure of indian antivenoms in Andaman and Nicobar Islands. *Frontiers in Pharmacology* 12, 768210.

Averin, A.S., Nenov, M.N., Starkov, V.G., Tsetlin, V.I., Utkin, Y.N., 2022. Effects of cardiotoxins from *Naja oxiana* cobra venom on rat heart muscle and aorta: A comparative study of toxin-induced contraction mechanisms. *Toxins* 14, 88.

Averin, A.S., Utkin, Y.N., 2021. Cardiovascular effects of snake toxins: Cardiotoxicity and cardioprotection. *Acta Naturae* 13, 4–14.

Bawaskar, H.S., Bawaskar, P.H., 2004. Envenoming by the common krait (*Bungarus caeruleus*) and Asian cobra (*Naja naja*): Clinical manifestations and their management in a rural setting. *Wilderness & Environmental Medicine* 15, 257–266.

Bawaskar, H.S., Bawaskar, P.H., Bawaskar, P.H., 2017. Snake bite in India: A neglected disease of poverty. *Lancet* 390, 1947–1948.

Bawaskar, H.S., Bawaskar, P.H., Punde, D.P., Inamdar, M.K., Dongare, R.B., Bhoite, R.R., 2008. Profile of snakebite envenoming in rural Maharashtra, India. *The Journal of the Association of Physicians of India* 56, 88–95.

Beraldo Neto, E., Coelho, G.R., Sciani, J.M., Pimenta, D.C., 2021. Proteomic characterization of *Naja mandalayensis* venom. *Journal of Venomous Animals and Toxins Including Tropical Diseases* 27, e20200125.

Buranasin, P., 1993. Snakebites at Maharat Nakhon Ratchasima Regional Hospital. *The Southeast Asian Journal of Tropical Medicine and Public Health* 24, 186–192.

Castro-Amorim, J., Novo de Oliveira, A., Da Silva, S.L., Soares, A.M., Mukherjee, A.K., Ramos, M.J., Fernandes, P.A., 2023. Catalytically active snake venom PLA(2) enzymes: An overview of its elusive mechanisms of reaction. *Journal of Medicinal Chemistry* 66, 5364–5376.

Castro, H.C., Zingali, R.B., Albuquerque, M.G., Pujol-Luz, M., Rodrigues, C.R., 2004. Snake venom thrombin-like enzymes: From reptilase to now. *Cellular and Molecular Life Sciences* 61, 843–856.

Chanda, A., Kalita, B., Patra, A., Senevirathne, W., Mukherjee, A.K., 2019. Proteomic analysis and antivenomics study of Western India *Naja naja* venom: Correlation between venom composition and clinical manifestations of cobra bite in this region. *Expert Review of Proteomics* 16, 171–184.

Chanda, A., Mukherjee, A.K., 2020. Quantitative proteomics to reveal the composition of Southern India spectacled cobra (*Naja naja*) venom and its immunological cross-reactivity towards commercial antivenom. *International Journal of Biological Macromolecules* 160, 224–232.

Chanda, A., Patra, A., Kalita, B., Mukherjee, A.K., 2018. Proteomics analysis to compare the venom composition between *Naja naja* and *Naja kaouthia* from the same geographical location of eastern India: Correlation with pathophysiology of envenomation and immunological cross-reactivity towards commercial polyantivenom. *Expert Review of Proteomics* 15, 949–961.

Chang, K.C., Huang, Y.K., Chen, Y.W., Chen, M.H., Tu, A.T., Chen, Y.C., 2020. Venom ophthalmia and ocular complications caused by snake venom. *Toxins* 12, 576.

Chen, K.C., Chiou, Y.L., Kao, P.H., Lin, S.R., Chang, L.S., 2008. Taiwan cobra cardiotoxins induce apoptotic death of human neuroblastoma SK-N-SH cells mediated by reactive oxygen species generation and mitochondrial depolarization. *Toxicon* 51, 624–634.

Chen, X., Lv, P., Liu, J., Xu, K., 2009. Apoptosis of human hepatocellular carcinoma cell (HepG2) induced by cardiotoxin III through S-phase arrest. *Experimental and Toxicologic Pathology* 61, 307–315.

Cher, C.D., Armugam, A., Zhu, Y.Z., Jeyaseelan, K., 2005. Molecular basis of cardiotoxicity upon cobra envenomation. *Cellular and Molecular Life Sciences* 62, 105–118.

Chu, E.R., Weinstein, S.A., White, J., Warrell, D.A., 2010. Venom ophthalmia caused by venoms of spitting elapid and other snakes: Report of ten cases with review of epidemiology, clinical features, pathophysiology and management. *Toxicon* 56, 259–272.

Das, D., Urs, N., Hiremath, V., Vishwanath, B.S., Doley, R., 2013. Biochemical and biological characterization of *Naja kaouthia* venom from North-East India and its neutralization by polyvalent antivenom. *Journal of Venom Research* 4, 31–38.

Dhananjaya, B.L., Nataraju, A., Rajesh, R., Raghavendra Gowda, C.D., Sharath, B.K., Vishwanath, B.S., D'Souza, C.J., 2006. Anticoagulant effect of *Naja naja* venom 5'nucleotidase: Demonstration through the use of novel specific inhibitor, vanillic acid. *Toxicon* 48, 411–421.

Du, X.Y., Clemetson, K.J., 2002. Snake venom L-amino acid oxidases. *Toxicon* 40, 659–665.

Dubovskii, P.V., Konshina, A.G., Efremov, R.G., 2014. Cobra cardiotoxins: Membrane interactions and pharmacological potential. *Current Medicinal Chemistry* 21, 270–287.

Dutta, S., Chanda, A., Kalita, B., Islam, T., Patra, A., Mukherjee, A.K., 2017. Proteomic analysis to unravel the complex venom proteome of eastern India *Naja naja*: Correlation of venom composition with its biochemical and pharmacological properties. *Journal of Proteomics* 156, 29–39.

Dutta, S., Gogoi, D., Mukherjee, A.K., 2015. Anticoagulant mechanism and platelet deaggregation property of a non-cytotoxic, acidic phospholipase A2 purified from Indian cobra (*Naja naja*) venom: Inhibition of anticoagulant activity by low molecular weight heparin. *Biochimie* 110, 93–106.

Dutta, S., Sinha, A., Dasgupta, S., Mukherjee, A.K., 2019. Binding of a *Naja naja* venom acidic phospholipase A(2) cognate complex to membrane-bound vimentin of rat L6 cells: Implications in cobra venom-induced cytotoxicity. *Biochimica et Biophysica Acta, Biomembranes* 1861, 958–977.

Escalante, T., Ortiz, N., Rucavado, A., Sanchez, E.F., Richardson, M., Fox, J.W., Gutierrez, J.M., 2011. Role of collagens and perlecan in microvascular stability: Exploring the mechanism of capillary vessel damage by snake venom metalloproteinases. *PLoS One* 6, e28017.

Estevao-Costa, M.I., Gontijo, S.S., Correia, B.L., Yarleque, A., Vivas-Ruiz, D., Rodrigues, E., Chávez-Olortegui, C., Oliveira, L.S., Sanchez, E.F., 2016. Neutralization of toxicological activities of medically-relevant Bothrops snake venoms and relevant toxins by two polyvalent bothropic antivenoms produced in Peru and Brazil. *Toxicon* 122, 67–77.

Faiz, M.A., Ahsan, M.F., Ghose, A., Rahman, M.R., Amin, R., Hossain, M., Tareq, M.N.U., Jalil, M.A., Kuch, U., Theakston, R.D.G., Warrell, D.A., Harris, J.B., 2017. Bites by the monocled cobra, *Naja kaouthia*, in Chittagong Division, Bangladesh: Epidemiology, clinical features of envenoming and management of 70 identified cases. *The American Journal of Tropical Medicine and Hygiene* 96, 876–884.

Gasanov, S.E., Shrivastava, I.H., Israilov, F.S., Kim, A.A., Rylova, K.A., Zhang, B., Dagda, R.K., 2015. *Naja naja oxiana* cobra venom cytotoxins CTI and CTII disrupt mitochondrial membrane integrity: Implications for basic three-fingered cytotoxins. *PLoS One* 10, e0129248.

Gutierrez, J.M., Calvete, J.J., Habib, A.G., Harrison, R.A., Williams, D.J., Warrell, D.A., 2017. Snakebite envenoming. *Nature Reviews. Disease Primers* 3, 17079.

Gutiérrez, J.M., Escalante, T., Rucavado, A., Herrera, C., 2016. Hemorrhage caused by snake venom metalloproteinases: A journey of discovery and understanding. *Toxins* 8, 93.

Gutierrez, J.M., Rucavado, A., 2000. Snake venom metalloproteinases: Their role in the pathogenesis of local tissue damage. *Biochimie* 82, 841–850.

Gutierrez, J.M., Theakston, R.D., Warrell, D.A., 2006. Confronting the neglected problem of snake bite envenoming: The need for a global partnership. *PLoS Medicine* 3, e150.

Hamza, M., Knudsen, C., Gnanathasan, C.A., Monteiro, W., Lewin, M.R., Laustsen, A.H., Habib, A.G., 2021. Clinical management of snakebite envenoming: Future perspectives. *Toxicon*, X(11), 100079.

Harris, R.J., Zdenek, C.N., Nouwens, A., Sweeney, C., Dunstan, N., Fry, B.G., 2020. A symmetry or asymmetry: Functional and compositional comparison of venom from the left and right glands of the Indochinese spitting cobra (*Naja siamensis*). *Toxicon*, X(7), 100050.

Harvey, A.L., Hider, R.C., Khader, F., 1983. Effect of phospholipase A on actions of cobra venom cardiotoxins on erythrocytes and skeletal muscle. *Biochimica et Biophysica Acta* 728, 215–221.

Hayes, W.K., Herbert, S.S., Rehling, G.C., Gennaro, J.F., 2002. Factors that influence venom expenditure in viperids and other snake species during predatory and defensive contexts. *Biology of the Vipers*, Eagle Mountain Publishing: Eagle Mountain, UT, USA, pp. 207–233.

Huang, H.W., Liu, B.S., Chien, K.Y., Chiang, L.C., Huang, S.Y., Sung, W.C., Wu, W.G., 2015. Cobra venom proteome and glycome determined from individual snakes of *Naja atra* reveal medically important dynamic range and systematic geographic variation. *Journal of Proteomics* 128, 92–104.

Hung, D.Z., Liau, M.Y., Lin-Shiau, S.Y., 2003. The clinical significance of venom detection in patients of cobra snakebite. *Toxicon* 41, 409–415.

Jagadeesha, D.K., Shashidhara Murthy, R., Girish, K.S., Kemparaju, K., 2002. A non-toxic anticoagulant metalloprotease: Purification and characterization from Indian cobra (*Naja naja naja*) venom. *Toxicon* 40, 667–675.

Kakati, H., Giri, S., Patra, A., Taye, S.J., Agarwalla, D., Boruah, H., Choudhary, G., Kalita, B., Mukherjee, A.K., 2023. A retrospective analysis of epidemiology, clinical features of envenomation, and in-patient management of snakebites in a model secondary hospital of Assam, North-east India. *Toxicon* 230, 107175.

Kakati, H., Patra, A., Kalita, B., Chanda, A., Rapole, S., Mukherjee, A.K., 2022. A comparison of two different analytical workflows to determine the venom proteome composition of *Naja kaouthia* from North-East India and immunological profiling of venom against commercial antivenoms. *International Journal of Biological Macromolecules* 208, 275–287.

Kalita, B., Utkin, Y.N., Mukherjee, A.K., 2022. Current insights in the mechanisms of cobra venom cytotoxins and their complexes in inducing toxicity: Implications in antivenom therapy. *Toxins* 14, 839.

Kao, P.H., Chen, K.C., Lin, S.R., Chang, L.S., 2008. The structural and functional contribution of N-terminal region and His-47 on Taiwan cobra phospholipase A2. *Journal of Peptide Science : An Official Publication of the European Peptide Society* 14, 342–348.

Kao, P.H., Wu, M.J., Chang, L.S., 2009. Membrane-bound conformation of *Naja nigricollis* toxin gamma affects its membrane-damaging activity. *Toxicon* 53, 342–348.

Kazandjian, T.D., Petras, D., Robinson, S.D., van Thiel, J., Greene, H.W., Arbuckle, K., Barlow, A., Carter, D.A., Wouters, R.M., Whiteley, G., Wagstaff, S.C., Arias, A.S., Albulescu, L.O., Plettenberg Laing, A., Hall, C., Heap, A., Penrhyn-Lowe, S., McCabe, C.V., Ainsworth, S., da Silva, R.R., Dorrestein, P.C., Richardson, M.K., Gutierrez, J.M., Calvete, J.J., Harrison, R.A., Vetter, I., Undheim, E.A.B., Wuster, W., Casewell, N.R., 2021. Convergent evolution of pain-inducing defensive venom components in spitting cobras. *Science* 371, 386–390.

Kini, R.M., Doley, R., 2010. Structure, function and evolution of three-finger toxins: Mini proteins with multiple targets. *Toxicon* 56, 855–867.

Konshina, A.G., Dubovskii, P.V., Efremov, R.G., 2012. Structure and dynamics of cardiotoxins. *Current Protein & Peptide Science* 13, 570–584.

Kularatne, S.A., Budagoda, B.D., Gawarammana, I.B., Kularatne, W.K., 2009. Epidemiology, clinical profile and management issues of cobra (*Naja naja*) bites in Sri Lanka: First authenticated case series. *Transactions of the Royal Society of Tropical Medicine and Hygiene* 103, 924–930.

Kwan, C.Y., Kwan, T.K., Huang, S.J., 2002. Effect of calcium on the vascular contraction induced by cobra venom cardiotoxin. *Clinical and Experimental Pharmacology & Physiology* 29, 823–828.

Laustsen, A.H., Gutierrez, J.M., Lohse, B., Rasmussen, A.R., Fernandez, J., Milbo, C., Lomonte, B., 2015. Snake venomics of monocled cobra (*Naja kaouthia*) and investigation of human IgG response against venom toxins. *Toxicon* 99, 23–35.

Liu, C.C., You, C.H., Wang, P.J., Yu, J.S., Huang, G.J., Liu, C.H., Hsieh, W.C., Lin, C.C., 2017. Analysis of the efficacy of Taiwanese freeze-dried neurotoxic antivenom against *Naja kaouthia, Naja siamensis* and *Ophiophagus hannah* through proteomics and animal model approaches. *PLoS Neglected Tropical Diseases* 11, e0006138.

Lomonte, B., Angulo, Y., Calderon, L., 2003. An overview of lysine-49 phospholipase A2 myotoxins from crotalid snake venoms and their structural determinants of myotoxic action. *Toxicon* 42, 885–901.

Mao, Y.C., Liu, P.Y., Chiang, L.C., Lai, C.S., Lai, K.L., Ho, C.H., Wang, T.H., Yang, C.C., 2018. *Naja atra* snakebite in Taiwan. *Clinical Toxicology* 56, 273–280.

McCue, M.D., 2006. Cost of producing venom in three North American pitviper species. *Copeia* 2006, 818–825.

Morgenstern, D., King, G.F., 2013. The venom optimization hypothesis revisited. *Toxicon* 63, 120–128.

Mukherjee, A., Ghosal, S., Maity, C., 2000. Some biochemical properties of Russell's viper (*Daboia russelli*) venom from Eastern India: Correlation with clinico-pathological manifestation in Russell's viper bite. *Toxicon* 38, 163–175.

Mukherjee, A.K., 2008. Characterization of a novel pro-coagulant metalloprotease (RVBCMP) possessing alpha-fibrinogenase and tissue haemorrhagic activity from venom of *Daboia russelli russelli* (Russell's viper): Evidence of distinct coagulant and haemorrhagic sites in RVBCMP. *Toxicon* 51, 923–933.

Mukherjee, A.K., 2010. Non-covalent interaction of phospholipase A(2) (PLA(2)) and kaouthiotoxin (KTX) from venom of *Naja kaouthia* exhibits marked synergism to potentiate their cytotoxicity on target cells. *Journal of Venom Research* 1, 37–42.

Mukherjee, A.K., 2014. The pro-coagulant fibrinogenolytic serine protease isoenzymes purified from *Daboia russelii russelii* venom coagulate the blood through factor V activation: Role of glycosylation on enzymatic activity. *PLoS One* 9, e86823.

Mukherjee, A.K., Mackessy, S.P., 2013. Biochemical and pharmacological properties of a new thrombin-like serine protease (Russelobin) from the venom of Russell's Viper (*Daboia russelii russelii*) and assessment of its therapeutic potential. *Biochimica et Biophysica Acta* 1830, 3476–3488.

Mukherjee, A.K., Maity, C.R., 2002. Biochemical composition, lethality and pathophysiology of venom from two cobras—*Naja naja* and *N. kaouthia*. *Comparative Biochemistry and Physiology. Part B, Biochemistry & Molecular Biology* 131, 125–132.

Mukherjee, A.K., Saviola, A.J., Burns, P.D., Mackessy, S.P., 2015. Apoptosis induction in human breast cancer (MCF-7) cells by a novel venom L-amino acid oxidase (Rusvinoxidase) is independent of its enzymatic activity and is accompanied by caspase-7 activation and reactive oxygen species production. *Apoptosis* 20, 1358–1372.

Mukherjee, A.K., Saviola, A.J., Mackessy, S.P., 2018. Cellular mechanism of resistance of human colorectal adenocarcinoma cells against apoptosis-induction by Russell's Viper venom l-amino acid oxidase (Rusvinoxidase). *Biochimie* 150, 8–15.

Nys, M., Zarkadas, E., Brams, M., Mehregan, A., Kambara, K., Kool, J., Casewell, N.R., Bertrand, D., Baenziger, J.E., Nury, H., Ulens, C., 2022. The molecular mechanism of snake short-chain alpha-neurotoxin binding to muscle-type nicotinic acetylcholine receptors. *Nature Communications* 13, 4543.

Palasuberniam, P., Chan, Y.W., Tan, K.Y., Tan, C.H., 2021. Snake venom proteomics of samar cobra (*Naja samarensis*) from the Southern Philippines: Short alpha-neurotoxins as the dominant lethal component weakly cross-neutralized by the Philippine cobra antivenom. *Frontiers in Pharmacology* 12, 727756.

Peterfi, O., Boda, F., Szabo, Z., Ferencz, E., Baba, L., 2019. Hypotensive snake venom components—A mini-review. *Molecules* 24.

Pucca, M.B., Knudsen, C., Oliveira, I.S., Rimbault, C., Cerni, F.A., Fan, Hui Wen, Sachett, J., Sartim, M.A., Laustsen, A.H., Monteiro, W.M., 2020. Current knowledge on snake dry bites. *Toxins* 12, 668.

Quyen, L., 2004. *Clinical evaluation of snakebites in Vietnam: A study from Cho Ray hospital.* Sch. Repos. National University of Singapore.

Ranawaka, U.K., Lalloo, D.G., de Silva, H.J., 2013. Neurotoxicity in snakebite—The limits of our knowledge. *PLoS Neglected Tropical Diseases* 7, e2302.

Raynor, R.L., Zheng, B., Kuo, J.F., 1991. Membrane interactions of amphiphilic polypeptides mastoparan, melittin, polymyxin B, and cardiotoxin. Differential inhibition of protein kinase C, Ca2+/calmodulin-dependent protein kinase II and synaptosomal membrane Na,K-ATPase, and Na+ pump and differentiation of HL60 cells. *The Journal of Biological Chemistry* 266, 2753–2758.

Sai Sein Lin, O., Myat Thet, N., Khin Maung, G., Than, A., Mi Mi, K., Myat Myat, T., Myo, T., Pyae Phyo, A., Oakkar Kyaw, K., Aye Zarchi, S., Du Wun, M., Htay, A., O'Shea, M., Mahmood, M.A., Peh, C.A., White, J., Warrell, D.A., 2020. Clinical importance of the Mandalay spitting cobra (*Naja mandalayensis*) in upper Myanmar—Bites, envenoming and ophthalmia. *Toxicon* 184, 39–47.

Saikia, D., Bordoloi, N.K., Chattopadhyay, P., Choklingam, S., Ghosh, S.S., Mukherjee, A.K., 2012. Differential mode of attack on membrane phospholipids by an acidic phospholipase A(2) (RVVA-PLA(2)-I) from *Daboia russelli* venom. *Biochimica et Biophysica Acta* 1818, 3149–3157.

Shan, L.L., Gao, J.F., Zhang, Y.X., Shen, S.S., He, Y., Wang, J., Ma, X.M., Ji, X., 2016. Proteomic characterization and comparison of venoms from two elapid snakes (*Bungarus multicinctus* and *Naja atra*) from China. *Journal of Proteomics* 138, 83–94.

Shashidharamurthy, R., Jagadeesha, D.K., Girish, K.S., Kemparaju, K., 2002. Variations in biochemical and pharmacological properties of Indian cobra (*Naja naja naja*) venom due to geographical distribution. *Molecular and Cellular Biochemistry* 229, 93–101.

Sintiprungrat, K., Watcharatanyatip, K., Senevirathne, W.D., Chaisuriya, P., Chokchaichamnankit, D., Srisomsap, C., Ratanabanangkoon, K., 2016. A comparative study of venomics of *Naja naja* from India and Sri Lanka, clinical manifestations and antivenomics of an Indian polyspecific antivenom. *Journal of Proteomics* 132, 131–143.

Slotta, K.H., Vick, J.A., 1969. Identification of the direct lytic factor from cobra venom as cardiotoxin. *Toxicon* 6, 167–173.

Slowinski, J.B., Wüster, W., 2000. A new cobra (Elapidae: Naja) from Myanmar (Burma). *Herpetologica*, 257–270.

Suntravat, M., Nuchprayoon, I., Perez, J.C., 2010. Comparative study of anticoagulant and procoagulant properties of 28 snake venoms from families Elapidae, Viperidae, and purified Russell's viper venom-factor X activator (RVV-X). *Toxicon* 56, 544–553.

Takeya, H., Nishida, S., Miyata, T., Kawada, S., Saisaka, Y., Morita, T., Iwanaga, S., 1992. Coagulation factor X activating enzyme from Russell's viper venom (RVV-X). A novel metalloproteinase with disintegrin (platelet aggregation inhibitor)-like and C-type lectin-like domains. *The Journal of Biological Chemistry* 267, 14109–14117.

Tan, C.H., Tan, K.Y., Wong, K.Y., Tan, N.H., Chong, H.P., 2022. Equatorial spitting cobra (*Naja sumatrana*) from Malaysia (Negeri Sembilan and Penang), Southern Thailand, and Sumatra: Comparative venom proteomics, immunoreactivity and cross-neutralization by antivenom. *Toxins* 14, 522.

Tan, C.H., Tan, N.H., 2015. Toxinology of snake venoms: The Malaysian context. *Snake Venoms*, 1–37.

Tan, C.H., Wong, K.Y., Chong, H.P., Tan, N.H., Tan, K.Y., 2019. Proteomic insights into short neurotoxin-driven, highly neurotoxic venom of Philippine cobra (*Naja philippinensis*) and toxicity correlation of cobra envenomation in Asia. *Journal of Proteomics* 206, 103418.

Tan, K.Y., Tan, C.H., Fung, S.Y., Tan, N.H., 2015. Venomics, lethality and neutralization of *Naja kaouthia* (monocled cobra) venoms from three different geographical regions of Southeast Asia. *Journal of Proteomics* 120, 105–125.

Tan, N.H., Wong, K.Y., Tan, C.H., 2017. Venomics of *Naja sputatrix*, the Javan spitting cobra: A short neurotoxin-driven venom needing improved antivenom neutralization. *Journal of Proteomics* 157, 18–32.

Tansuwannarat, P., Tongpoo, A., Phongsawad, S., Sriapha, C., Wananukul, W., Trakulsrichai, S., 2023. A retrospective cohort study of cobra envenomation: Clinical characteristics, treatments, and outcomes. *Toxins* 15, 468.

Teixeira, C., Cury, Y., Moreira, V., Picolob, G., Chaves, F., 2009. Inflammation induced by *Bothrops asper* venom. *Toxicon* 54, 988–997.

Thakur, R., Chattopadhyay, P., Ghosh, S.S., Mukherjee, A.K., 2015. Elucidation of procoagulant mechanism and pathophysiological significance of a new prothrombin activating metalloprotease purified from *Daboia russelii russelii* venom. *Toxicon* 100, 1–12.

Thakur, R., Mukherjee, A., 2015. A brief appraisal on Russell's viper venom (*Daboia russelii russelii*) proteinases. *Snake Venoms*, 1–18.

Tsai, T.H., Lin, C.C., Mao, Y.C., Hung, C.L., Yang, Y.C., Yang, C.C., Jeng, M.J., 2020. *Naja atra* venom-spit ophthalmia in Taiwan: An epidemiological survey from 1990 to 2016. *Journal of the Chinese Medical Association : JCMA* 83, 77–83.

Utkin, Y.N., 2013. Three-finger toxins, a deadly weapon of elapid venom—Milestones of discovery. *Toxicon* 62, 50–55.

Wang, C.H., Liu, J.H., Lee, S.C., Hsiao, C.D., Wu, W.G., 2006. Glycosphingolipid-facilitated membrane insertion and internalization of cobra cardiotoxin. The sulfatide.cardiotoxin complex structure in a membrane-like environment suggests a lipid-dependent cell-penetrating mechanism for membrane binding polypeptides. *The Journal of Biological Chemistry* 281, 656–667.

Wang, C.H., Wu, W.G., 2005. Amphiphilic beta-sheet cobra cardiotoxin targets mitochondria and disrupts its network. *FEBS Letters* 579, 3169–3174.

Watt, G., Padre, L., Tuazon, L., Theakston, R.D., Laughlin, L., 1988. Bites by the Philippine cobra (*Naja naja philippinensis*): Prominent neurotoxicity with minimal local signs. *The American Journal of Tropical Medicine and Hygiene* 39, 306–311.

Wong, K.Y., Tan, C.H., Tan, K.Y., Quraishi, N.H., Tan, N.H., 2018. Elucidating the biogeographical variation of the venom of *Naja naja* (spectacled cobra) from Pakistan through a venom-decomplexing proteomic study. *Journal of Proteomics* 175, 156–173.

Wong, K.Y., Tan, K.Y., Tan, N.H., Gnanathasan, C.A., Tan, C.H., 2021. Elucidating the venom diversity in Sri Lankan spectacled cobra (*Naja naja*) through De Novo Venom gland transcriptomics, venom proteomics and toxicity neutralization. *Toxins* 13, 558.

Wong, O.F., Lam, T.S., Fung, H.T., Choy, C.H., 2010. Five-year experience with Chinese cobra (*Naja atra*)--related injuries in two acute hospitals in Hong Kong. *Hong Kong Medical Journal* 16, 36–43.

Wongtongkam, N., Wilde, H., Sitthi-Amorn, C., Ratanabanangkoon, K., 2005. A study of Thai cobra (*Naja kaouthia*) bites in Thailand. *Military Medicine* 170, 336–341.

Wu, W.G., 1998. Cobra cardiotoxin and phospholipase A2 as GAG-binding toxins: On the path from structure to cardiotoxicity and inflammation. *Trends in Cardiovascular Medicine* 8, 270–278.

Xiao, G., Liu, J., Peng, L., Yang, Y., Sun, Z., 2022. Compositional and toxicological investigation of pooled venom from farm-raised *Naja atra*. *The Journal of Venomous Animals and Toxins Including Tropical Diseases* 28, e20210040.

Xu, N., Zhao, H.Y., Yin, Y., Shen, S.S., Shan, L.L., Chen, C.X., Zhang, Y.X., Gao, J.F., Ji, X., 2017. Combined venomics, antivenomics and venom gland transcriptome analysis of the monocoled cobra (*Naja kaouthia*) from China. *Journal of Proteomics* 159, 19–31.

Yang, S.H., Chien, C.M., Chang, L.S., Lin, S.R., 2007. Involvement of c-jun N-terminal kinase in G2/M arrest and caspase-mediated apoptosis induced by cardiotoxin III (*Naja naja atra*) in K562 leukemia cells. *Toxicon* 49, 966–974.

Yin, X., Guo, S., Gao, J., Luo, L., Liao, X., Li, M., Su, H., Huang, Z., Xu, J., Pei, J., Chen, S., 2020. Kinetic analysis of effects of temperature and time on the regulation of venom expression in *Bungarus multicinctus*. *Scientific Reports* 10, 14142.

Zhang, B., Li, F., Chen, Z., Shrivastava, I.H., Gasanoff, E.S., Dagda, R.K., 2019. *Naja mossambica mossambica* cobra cardiotoxin targets mitochondria to disrupt mitochondrial membrane structure and function. Toxins 11, 152.

Zhang, C., Medzihradszky, K.F., Sanchez, E.E., Basbaum, A.I., Julius, D., 2017. Lys49 myotoxin from the Brazilian lancehead pit viper elicits pain through regulated ATP release. *Proceedings of the National Academy of Sciences of the United States of America* 114, E2524–E2532.

Zhao, C., Zhao, J., Yang, Q., Ye, Y., 2017. Cobra neurotoxin produces central analgesic and hyperalgesic actions via adenosine A(1) and A(2A) receptors. *Molecular Pain* 13, 1744806917720336.

7 Treatment and Hospital Management of Cobra Envenomation in the Asian Continent

7.1 ASIAN COBRA ENVENOMATION: DETERMINANTS OF MORBIDITY AND MORTALITY OF BITE VICTIMS

Envenomation by snakebites, mainly a Cobra species, is a grave concern and is responsible for many mortalities and morbidities in the Asian continent. As described in the previous chapter, the severity of envenomation is dictated by the inflicting species, the amount of venom injected during envenomation, and the promptness of appropriate treatment. While the former two factors are intractable, transferring the snake-envenomed victims quickly to the nearest healthcare facility for immediate clinical assessment and treatment can save the lives of victims.

It has been well established in several snakebite epidemiological studies that a delay in the clinical assessment of the envenoming incident significantly increases the risk of snakebite-associated complications. Cobra venom toxins are fast-acting molecules; thus, delayed treatment can fuel the rapid dissemination of venom toxins in the victims' bodies to cause irreversible damage. Therefore, clinicians who treat snakebites suggest that clinical intervention within the first two hours (also known as the 'golden hour') is crucial for a good outcome for the snake-envenomed patients (Mukherjee and Mackessy, 2021).

Regrettably, snake envenomings, including those by the Asian Cobra species, mainly affect the rural and agricultural population in remote villages where specialized facilities for snakebite treatment are scarce. In addition, these remote areas are not well-connected to towns or cities and lack the necessary transportation facilities for transporting the envenomed victims to the closest points of care, causing a delay in clinical intervention. Another serious issue regarding the management of envenomed victims in the Asian continent, specifically in the rural areas, is the prevalence of superstitious beliefs among the rural dwellers who increasingly tend to consult local traditional healers instead of visiting a healthcare center (Mukherjee, 2021; Mukherjee and Mackessy, 2021; Patra and Mukherjee, 2021). Therefore, a delay between the time of envenomation and commencement of treatment significantly contributes to the mortality and morbidity of Asian Cobra envenomed victims.

DOI: 10.1201/9781003598121-7

7.2 DIAGNOSIS OF COBRA ENVENOMATION: AN ESSENTIAL FACTOR FOR SUCCESSFUL TREATMENT OF COBRA ENVENOMATION

The attending physician primarily decides the appropriate treatment modalities based on the presenting signs and symptoms of the victims and the identity of the inflicting snake species, if available (Mukherjee et al., 2021; Mukherjee and Mackessy, 2021; Puzari and Mukherjee, 2020). The clinical symptoms of snake envenomation can be broadly classified into neurotoxic and hemotoxic manifestations, requiring discrete clinical assessment and management. Therefore, determining the type of envenomation during the preliminary examination is necessary for better hospital management of snake-envenomed patients (Mukherjee, 2021; Mukherjee et al., 2021).

7.2.1 Clinical and Syndromic Diagnosis

Cobra envenomations are primarily associated with noticeable acute neuroparalysis within 30 to 60 min of envenomation, and therefore, the manifestation of neurotoxic symptoms by patients may hint at a Cobra bite. However, the scenario may get complicated if other snake species with neurotoxic venoms inhabit the geographic location of the envenomation incident. For example, apart from *N. naja* envenomation, *Daboia russelii* (a member of the Viperidae family) envenomed patients in Sri Lanka and southern India manifest neuroparalysis (Kalita et al., 2018; Silva et al., 2016; Suchithra et al., 2008). However, the hospital management of victims of *N. naja* and *D. russelii* envenomation is quite distinct since the latter species also manifests deranged coagulopathy, which is absent in Cobra envenomation. Therefore, correctly identifying the offending species is crucial for effective hospital management of snake-envenomed patients as it aids the attending physician in promptly starting the proper treatment. The above facts underscore the importance of dedicated Cobra envenomation detection kits for a confirmed diagnosis, which are unfortunately unavailable in the Asian continent.

Nevertheless, in some instances, the physician may predict the inflicting snake species based on the description given by the victim or any eyewitness of the incident. In some cases, the photographs may be brought with the patient, which can also help to ascertain the culprit species. There have also been instances where the live or dead specimen of the inflicting snake species was brought to the hospital along with the patient. In addition, the differences in the geographic distribution of a particular Cobra species can also be considered to predict the offending species. However, suppose the physician disagrees with the identity of the confronting species. In that case, they must rely on relevant experience of managing similar snake-envenomed patients (syndromic diagnosis).

7.2.2 Accurate Diagnosis of Cobra Envenomation: An Urgent Need for Venom Detection Kits

As discussed in the above section, similar clinical manifestations can be observed in patients envenomed by related or unrelated snake species, which may lead to misguided treatment modalities. Only in rare cases may the patient or accompanying individual bring the dead or alive snake specimen involved in the incident. Moreover, the patient's description of the inflicting species may not always be reliable since the victim would naturally be under tremendous shock and anxiety. Under these circumstances, identifying the confronting Cobra species using scientifically validated methods is vital for a prudent diagnosis. Unfortunately, Australia is the only country currently employing snake venom detection kits to identify the confronting species in clinical settings. The detection kit is based on a rapid two-step enzyme immunoassay that uses specific antibodies generated against snake venoms, which are medically relevant in Australia. However, such kits are not available to detect Cobra envenomation in Asia.

Several studies over the last two decades have shown the feasibility of developing such kits to identify some of the Cobra species of the Asian continent. However, they are still not commercialized, perhaps due to their high manufacturing cost, cross-reactivity with venoms of other related species, or assay time (Puzari and Mukherjee, 2020). A reverse latex agglutination-based assay was developed by Chinonavanig and co-workers that could specifically detect *N. kaouthia* venom (Thailand) toxins in wound swabs and plasma, serum, and urine samples of envenomed victims within 40 minutes. In addition, ELISA-based tests have also been developed for the detection of *N. atra* (Taiwan), *N. kaouthia* (Thailand), and *N. naja* (India) venoms with appreciable detection limits of ~1 ng/ml, 10 ng/ml, and 0.1 ng/ml, respectively. However, it was found that the anti-*N. naja* antibodies used in the ELISA could also cross-react with *Bungarus caeruleus* venoms, limiting its clinical application for accurate species identification (Chinonavanig et al., 1988).

Furthermore, a modified ELISA based on avidin-biotin reaction was developed to detect *N. naja* (southern Vietnam) venom in wound exudates of envenomed victims that could complete detection within 35 minutes with a detection limit of 0.2 ng/ml (Van Dong et al., 2003). Similarly, Liu et al. (2018) developed a sandwich ELISA-based assay to identify *N. atra* (Taiwan) venom in the plasma of experimentally envenomed mice within 30 minutes at a detection limit of 0.2 ng/ml (Liu et al., 2018). Dong and co-workers (2004) developed an optical immunoassay-based test for the detection of *N. kaouthia* (southern Vietnam) venom in whole blood, plasma, urine, and wound exudates of envenomed patients with detection limits of 0.2 to 200 ng/ml and assay time of 33 minutes (Dong et al., 2004). Application of PCR-based amplification of cobrotoxin encoding gene has also been demonstrated to detect *N. kaouthia* (Thailand) envenomation in mice; however, the assay takes 3 to 4 hours to complete, thereby making it a relatively less attractive candidate for development of Cobra envenomation detection kit (Suntrarachun et al., 2001). Unfortunately, none of these assays have been routinely deployed to identify Cobra envenomation in clinical settings. Therefore, considering the severity and frequency of Cobra envenomation in the Asian continent, the development and clinical application of Cobra venom detection kits that are rapid, species-specific, sensitive, and inexpensive is of utmost importance for the successful management of Cobra envenomation.

7.3 TREATMENT MODALITIES OF COBRA ENVENOMATION

The infrastructure and personnel expertise available at respective healthcare centers influences the treatment modality (Figure 7.1). Furthermore, the treatment of snake envenomation may vary on a case-to-case basis, albeit they must be undertaken with utmost urgency. Irrespective of the Cobra (or any snake) species involved in an envenomation incident, a thorough clinical assessment of the patient is done to determine the severity of the envenomation. To begin with, the clinician checks for any fang marks and local effects at the bite site, along with assessing the blood pressure, heart rate, respiratory rate, power of neck flexors and limb muscles, and the level of consciousness of the patient. In addition, the clinician constantly evaluates the patient for the progression of neuroparalytic symptoms, including blurring of vision, dysarthria, and ptosis, to confirm a neurotoxic envenomation. The clinician can also rely on a negative 20-minute whole blood clotting test (20WBCT) to rule out hemotoxic envenoming (Puzari and Mukherjee, 2020).

The following sub-sections highlight the different treatment choices frequently employed for hospital management of Cobra envenomation throughout Asia. However, it is noteworthy that while some of the treatment modalities described below are mandatory for any confirmed Cobra-envenomed victims, others might be necessary depending on the severity of the envenomation and the recovery of the patients.

7.3.1 First-Aid Measures

First-aid measures aim to minimize the systemic dissemination of venom toxins and prevent the victims' life-threatening complications. As a result, these measures should be started almost immediately after the envenomation incident, either by the

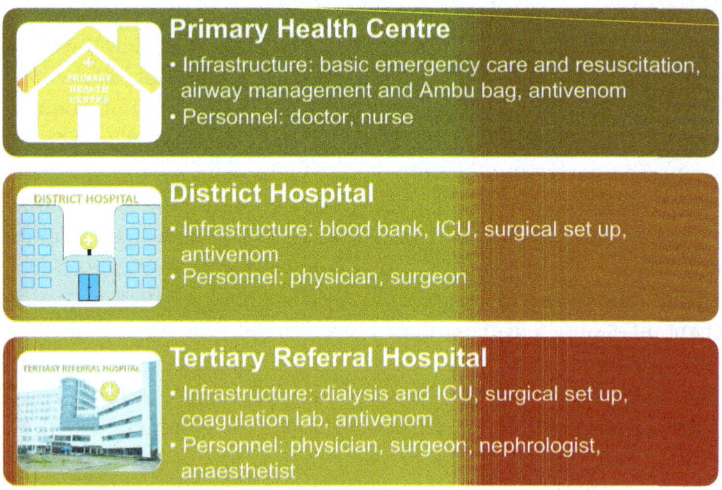

FIGURE 7.1 A diagram showing the different types of healthcare centers for treating snakebites and the respective infrastructure available in these establishments.

victim or by any individual who is present nearby. Naturally, the victim will develop immense fear and shock due to the unforeseen episode. Therefore, emphasis must be placed on reassuring the patient that there is no reason to be nervous or frightened. Moreover, care must be taken to immobilize the victim's affected limb to prevent the spread of venom. These actions must be followed by promptly transferring the victim to a hospital where additional first-aid measures can be provided (Mukherjee and Mackessy, 2021).

The healthcare center's first-aid measures constitute local wound care, analgesics, and sterile dressing. In addition, since the mouth of snakes, including Cobras, might contain an array of aerobic as well as anaerobic bacteria, a prophylactic antibiotic (penicillin or erythromycin for penicillin-hypersensitive patients) course along with a single broad-spectrum antibiotic dose (to counter anaerobic species) is always preferred. Moreover, irrespective of the severity of envenomation, tetanus prophylaxis is also used to prevent infection of the Cobra bite wound (WHO, 2016).

7.3.2 Antivenom Therapy

Antivenoms remain the mainstay and the only World Health Organization (WHO)-approved treatment for Cobra envenomation. They are produced in large animals, such as horses, donkeys, sheep, or camels. However, horses are preferred for the commercial production of antivenoms due to their robust immune system and higher blood volumes (Ratanabanangkoon, 2023). Antivenoms contain whole IgG molecules or pepsin-digested $F(ab')_2$ fragments that have been raised either against a single Cobra venom (monovalent) or against two snake species, among which one is a Cobra species (bivalent) or a mixture of medically relevant snake venoms, including those of Cobras in a discrete ratio (polyvalent). For example, Queen Saovabha Memorial Institute, Thailand, manufactures monovalent antivenom (MAV) by immunizing horses with sub-lethal doses of *N. kaouthia* (Thailand) venoms. Notably, a commercial monovalent antivenom against Indian *N. kaouthia* venom was produced by VINS Bioproducts Limited, India, but the product has now been discontinued. The Taiwanese Neuro Bivalent Antivenom (NBAV) manufactured by Taiwan Central for Disease Control is a bivalent antivenom raised against Taiwanese *Bungarus multicintus* and *N. atra* venoms. On the contrary, the immunization mixture for the production of Indian polyvalent antivenom (PAV) contains venoms of 'Big Four' snake species, namely *N. naja, Bungarus caeruleus* (Indian common krait), *Daboia russelii* (Indian Russell's Viper), and *Echis carinatus* (Indian Saw-scaled viper).

Since MAV contains specific antibodies against a single Cobra species, it is logical that MAV therapy is a better treatment choice over PAV if the identity of the inflicting Cobra species is confirmed. However, as in most cases, the offending species cannot be identified with surety, physicians usually prefer to treat Cobra-envenomed patients with PAV. Irrespective of the type of antivenom, the antibodies present in the antivenom preparation aim to recognize and thereby neutralize the venom toxins when administered to the patients. The antivenoms that are commercially available and are routinely used for the management of Asian Cobra envenomation are listed in Table 7.1.

TABLE 7.1
List of Antivenoms Available for Treating Cobra Envenomation in Different Asian Countries

Cobra Species	Region	Antivenom Name	Manufacturer	Antivenom Type (Antibody Format)	Specificity	Venom Immunogen Used for the Antivenom Production
N. atra	China	*Naja naja (atra)* Antivenin	Shanghai Serum Bio-technology Co., Ltd., China	Monovalent (IgG)	Specific	*N. atra*
	Taiwan	Freeze-dried neurotoxic antivenom (FNAV)	Taiwan Centers for Disease Control, Taiwan	Bivalent (IgG)	Specific	*N. atra and Bungarus multicinctus*
N. kaouthia	Vietnam	SAV-*Naja*	Institute of Vaccines and Medical Biologicals, Vietnam	Monovalent (IgG)	Specific	*N. kaouthia*
		Neuro Polyvalent Snake Antivenin	Queen Saovabha Memorial Institute, Thailand	Polyvalent (F(ab')$_2$)	Specific	*Ophiophagus hannah, N. kaouthia, B. fasciatus, B. candidus*
	Myanmar	Cobra antivenom	Burma Pharmaceutical Industries, Myanmar	Monovalent (IgG)	Specific	*N. kaouthia*
	Thailand	Cobra Antivenin	Queen Saovabha Memorial Institute, Thailand	Monovalent (F(ab')$_2$)	Specific	*N. kaouthia*
		Neuro Polyvalent Snake Antivenin	Queen Saovabha Memorial Institute, Thailand	Polyvalent (F(ab')$_2$)	Specific	*Ophiophagus hannah, N. kaouthia, B. fasciatus, B. candidus*
	Malaysia	Cobra Antivenin	Queen Saovabha Memorial Institute, Thailand	Monovalent (F(ab')$_2$)	Specific	*N. kaouthia*
		Neuro Polyvalent Snake Antivenin	Queen Saovabha Memorial Institute, Thailand	Polyvalent (F(ab')$_2$)	Specific	*Ophiophagus hannah, N. kaouthia, B. fasciatus, B. candidus*

	India	Snake antivenin I.P.	Haffkine Biopharmaceutical Corporation Ltd., India	Polyvalent (F(ab')$_2$)	Paraspecific	*N. naja, B. caeruleus, Daboia russelii, Echis carinatus*
		Polyvalent Snake Antivenin	Biological E Limited, India	Polyvalent (F(ab')$_2$)	Paraspecific	-do-
		Snake Venom Antiserum (Polyvalent)	Bharat Serums and Vaccines Limited, India	Polyvalent (F(ab')$_2$)	Paraspecific	-do-
		Polyvalent Anti Snake Venom Serum I.P.	King Institute of Preventative Medicine and Research, India	Polyvalent (F(ab')$_2$)	Paraspecific	-do-
		Snake Venom Antiserum I.P.	VINS Bioproducts Ltd., India	Polyvalent (F(ab')$_2$)	Paraspecific	-do-
		Snake venom antiserum I.P.	Premium Serums and Vaccines Pvt. Ltd., India	Polyvalent (F(ab')$_2$)	Paraspecific	-do-
N. naja	India, Sri Lanka, Nepal, Bangladesh	Snake antivenin I.P.	Haffkine Biopharmaceutical Corporation Ltd., India	Polyvalent (F(ab')$_2$)	Specific	-do-
		Polyvalent Snake Antivenin	Biological E Limited, India	Polyvalent (F(ab')$_2$)	Specific	-do-
		Snake Venom Antiserum (Polyvalent)	Bharat Serums and Vaccines Limited, India	Polyvalent (F(ab')$_2$)	Specific	-do-
		Polyvalent Anti Snake Venom Serum I.P.	King Institute of Preventative Medicine and Research, India	Polyvalent (F(ab')$_2$)	Specific	-do-
		Snake Venom Antiserum I.P.	VINS Bioproducts Ltd., India	Polyvalent (F(ab')$_2$)	Specific	-do-
		Snake venom antiserum I.P.	Premium Serums and Vaccines Pvt. Ltd., India	Polyvalent (F(ab')$_2$)	Specific	-do-
	Pakistan	Polyvalent Antisnake Venom Serum	National Institute of Health, Pakistan	Polyvalent (not mentioned)	Specific	-do-

(Continued)

TABLE 7.1 (CONTINUED)
List of Antivenoms Available for Treating Cobra Envenomation in Different Asian Countries

Cobra Species	Region	Antivenom Name	Manufacturer	Antivenom Type (Antibody Format)	Specificity	Venom Immunogen Used for the Antivenom Production
N. oxiana	Iran	Hexavalent snake venom hexavalent immunoglobulin	Razi Vaccine & Serum Research Institute, Iran	Polyvalent (IgG)	Specific	*N. oxiana, Pseudocerastes persicus fieldi, Echis carinatus, Vipera albicornuta, Vipera lebetina obtusa, and Agkistrodon halys*
		SnaFab6	Padra Serum Alborz, Iran	Polyvalent (F(ab')$_2$)	Specific	-do-
		Polyvalent Antisnake Venom Serum	National Institute of Health, Pakistan	Polyvalent (not mentioned)	Paraspecific	*N. naja, B. caeruleus, Daboia russelii, Echis carinatus*
N. philippinensis	Philippines	Purified Cobra Antivenom	Research Institute of Tropical Medicine, Philippines	Monovalent (IgG)	Specific	*N. philippinensis*
N. samarensis	Philippines	Purified Cobra Antivenom	Research Institute of Tropical Medicine, Philippines	Monovalent (IgG)	Paraspecific	*N. philippinensis*
N. siamensis	Thailand	Cobra Antivenin	Queen Saovabha Memorial Institute, Thailand	Monovalent (F(ab')$_2$)	Paraspecific	*N. kaouthia*
		Neuro Polyvalent Snake Antivenin	Queen Saovabha Memorial Institute, Thailand	Polyvalent (F(ab')$_2$)	Paraspecific	*Ophiophagus hannah, N. kaouthia, B. fasciatus, B. candidus*
N. sputatrix	Java Island	Serum Anti Bisa Ular (BioSAVE)	Biofarma, Indonesia	Polyvalent (F(ab')$_2$)	Specific	*N. sputatrix, Bungarus fasciatus, and Agkistrodon rhodostoma*

N. sumatrana	Thailand	Cobra Antivenin	Queen Saovabha Memorial Institute, Thailand	Monovalent ($F(ab')_2$)	Paraspecific	*N. kaouthia*
		Neuro Polyvalent Snake Antivenin	Queen Saovabha Memorial Institute, Thailand	Polyvalent ($F(ab')_2$)	Paraspecific	*Ophiophagus hannah, N. kaouthia, B. fasciatus, B. candidus*
N. sagittifera	India	Currently, no antivenom is available, and there is poor para specificity against Indian polyvalent antivenom (Bharat Serums and Vaccines Limited; Premium Serums and Vaccines Pvt. Ltd.)				
N. mandalayensis	Myanmar	Currently, no antivenom is available, and there is poor para specificity against Cobra antivenom (Burma Pharmaceutical Industries) and Indian polyvalent antivenom.				

Notes: The data has been retrieved from the Snakebite Information and Data Platform (World Health Organization).

Specific commercial antisnake antivenoms are only available against six of the eleven Asian Cobra species – *N. atra, N. kaouthia, N. naja, N. oxiana, N. sputatrix* and *N. philippinensis*. Among them, MAV is available for *N. atra* (China and Taiwan), *N. kaouthia* (Myanmar, Thailand, Malaysia, and Vietnam), and *N. philippinensis* (Philippines), while patients envenomed by the rest of the three species are treated by administering PAV. Notably, in India, the treatment of *N. kaouthia* envenomation is usually done using a paraspecific polyvalent antivenom (mentioned above) because specific MAV/PAV against this species is not available in this country.

Similarly, the treatment of envenomation by three other Cobra species, namely, *N. samarensis, N. siamensis*, and *N. sumatrana*, is also performed using polyvalent or monovalent paraspecific antivenoms that are raised against other Cobra species (Table 7.1). Paraspecificity is a phenomenon in which an antivenom raised against the venom of a particular Cobra species can recognize and thereby neutralize (at least partially) venoms of a closely related Cobra species due to certain similarities in their composition. Although para specificity has been established in certain antivenoms (Attarde et al., 2021; Chanda et al., 2018; Kakati et al., 2022; Palasuberniam et al., 2021; Tan et al., 2022), it is expected that the administration of congeneric antivenoms can only provide limited protection. Therefore, treatment should commence with a specific antivenom whenever it is available. The WHO also recommends using paraspecific antivenoms only when a specific antivenom is unavailable for a particular species (WHO, 2016). Nevertheless, the paraspecificity of antivenoms must be thoroughly investigated and proved under in vitro and in vivo conditions before their clinical utilization.

Unfortunately, specific antivenoms for the treatment of envenomation by *N. sagittifera* and *N. mandalayensis* are currently not commercially available. Moreover, a couple of studies have shown that these Cobra venoms exhibit poor paraspecificity against Indian polyvalent antivenom (manufactured by Bharat Serums and Vaccines Limited and Premium Serums and Vaccines Pvt. Ltd.) and Cobra antivenom (manufactured by Burma Pharmaceutical Industries), thereby making hospital management of envenomation by these two Cobra species, particularly challenging (Attarde et al., 2021; Sai Sein Lin et al., 2020).

Apart from the availability of specific or at least paraspecific antivenom, another important consideration for managing confirmed/suspected Cobra envenomed patients is whether antivenom administration is necessary. As discussed in Chapter 6, Cobras (20 to 30% of cases) frequently deliver dry bites without injecting venom into the victims. Therefore, administering antivenom in such scenarios might be detrimental to the patient instead of recovery because antivenoms contain IgG or $F(ab')_2$ molecules that can induce early anaphylactic, late (serum sickness-type), and pyrogenic reactions (see Section 7.3.6 for details). Therefore, antivenom should be administered when the benefits for the patient exceed the risks (Seneviratne et al., 2000; WHO, 2016).

To manage neurotoxic envenoming, including Cobra bites, WHO guidelines recommend treating neurotoxic envenomation by antivenom when signs of neurotoxicity such as bilateral ptosis, external ophthalmoplegia, and paralysis are recorded by the attending physician. Moreover, for Asian Cobra envenoming cases with predominant local tissue damage (as in the case of *N. atra* and *N. siamensis*), indications of antivenom administration include persistent local swelling in more than half of the bitten limb within 48 hours of envenomation or rapid extension of swelling within few hours of envenomation. However, patients exhibiting mild symptoms, including pain at the bite site and local swelling without neurological manifestation, can be treated conservatively.

The initial and follow-up (if required) doses of antivenom administration are also crucial for favorable outcomes of Cobra-envenomed patients. However, clinical trial data on managing snake-envenomed patients with variable severities is hardly available. Further, the antivenom manufacturers' information is based only on LD_{50} and ED_{50} values determined in mouse models and, thus, are clinically unreliable. Therefore, the selection of starting antivenom dose is generally empirical. It has been found across several epidemiological reports that a higher initial dose of antivenom is beneficial compared to the administration of lower doses over several days. For hospital management of Cobra envenomation, an initial dose containing 5–10 vials of antivenom is administered in most cases. However, in some cases, fewer MAV vials (<5) have also been used as an initial dose for treating mild envenoming by *N. atra* in Hong Kong (WHO, 2016). This fact highlights that specific MAV therapy is a better treatment choice over PAV (or paraspecific PAV/MAV) if the identity of the confronting Cobra species is known.

On the contrary, in a few cases of severe *N. kaouthia* envenoming (n = 19) in northeast India, administering an initial dose of 20 paraspecific antivenom (PAV against Big Four) vials has also proven successful. It was found that the acute neurotoxic symptoms of envenomation were typically reversed within 1–2 hours after the commencement of antivenom therapy (Kakati et al., 2023). After the initial dose, the patients are closely monitored for at least 1 hour, and if the neurotoxic symptoms persist or worsen, a follow-up dose of 5–10 vials is administered (Mukherjee et al., 2021; WHO, 2016). However, repetition of follow-up doses is usually avoided if the patient is paralyzed and under mechanical ventilation. Antivenoms are typically administered intravenously as a slow IV injection with a maximal delivery rate of 2 ml per minute or are diluted in an isotonic fluid (5 ml/kg) and administered for 30–60 minutes (Mukherjee et al., 2021; WHO, 2016). Other routes of antivenom administration, including intramuscular injection, are not very effective as they result in poor bioavailability of the antibodies and thus offer minimal benefit to the patients (WHO, 2016).

7.3.3 Supportive Treatment

Apart from antivenom therapy, acetylcholinesterase (AChE) inhibitor administration has also proved beneficial for alleviating neurological manifestations in Cobra-envenomed patients. The enzyme AChE is responsible for the rapid hydrolysis of acetylcholine neurotransmitters, contributing to neurological manifestations in envenomed victims (Ranawaka et al., 2013). Therefore, an inhibitor of this enzyme prevents hydrolysis of acetylcholine, thereby increasing the level and duration of the neurotransmitter action. The above treatment, in turn, aids in the restoration of some of the neurotoxic symptoms in the treated patients.

The WHO has approved two classes of AChE inhibitors for Cobra envenomed patients with neurological manifestations – the fast-acting edrophonium chloride

(Tensilon) and the relatively slow-acting neostigmine (or distigmine, pyridostigmine, ambenomium, etc.). The recommended dose of the fast-acting inhibitor is 10 mg/kg for adults or 0.25 mg/kg for children, and it is usually administered as a slow intravenous injection (WHO, 2016). However, it is essential to note that to counter the risk of life-threatening cholinergic crisis (bradycardia, bronchoconstriction, bronchorrhea, alimentary peristalsis, etc.) of AChE inhibitors, the patients must be administered appropriate doses of anticholinergic drugs before administering AChE inhibitors.

In the case of neostigmine and its derivatives, the approved dose is 0.02 mg/kg for adults and 0.04 mg/kg for children, and they can be administered either intramuscularly or intravenously. The WHO recommends intravenous administration of atropine sulfate (at a dose of 0.6 mg/kg in adults or 0.05 mg/kg in children) or glycopyrronium (glycopyrrolate) (amount not specified) in case of patients undergoing AChE inhibitor treatment. These recommended doses of AChE inhibitors in conjunction with anticholinergic drugs have been widely used to manage the neurotoxic symptoms of Cobra-envenomed patients in Asia (WHO, 2016).

Although these temporizing drugs benefit most of the Cobra envenomed cases with neurotoxicity, the patients must be closely monitored for a positive response before continuing with additional doses. The patients administered with neostigmine are monitored for 30 to 60 minutes, and those given edrophonium chloride are observed for 10–20 minutes for any visible signs of improved neuromuscular function. The patients that respond positively to the AChE inhibitor and anticholinergic drug combination therapy can be given additional doses of neostigmine at a rate of 0.5–2.5 mg every 1–3 hours for adults or 0.01–0.04 mg/kg every 2 –4 hours; the maximum permissible dose being 10 mg/24 hours. As in the case of the initial amount, the follow-up doses of AChE inhibitor should also be preceded by the administration of atropine sulfate or glycopyrronium (WHO, 2016).

Cobra bites often result in extreme pain at the bite site; therefore, physicians rely on analgesic drugs to alleviate the pain of such patients. WHO-recommended doses of analgesics include acetaminophen (paracetamol) at doses of 500 mg to 1 g/day in adults (maximum dose 4 g/adult/day) or 10–15 mg/kg/day in children (maximum dose 100 mg/kg/day) and codeine phosphate at doses of 30–60 mg/day in adults (maximum dose 240 mg/adult/day) or 0.5 mg/kg/day in children of more than two years of age (maximum dose 2 mg/kg/day). The use of codeine phosphate in children under two years of age is not approved. These analgesics are usually administered orally for 4 to 6 hours until the pain subsides (WHO, 2016).

7.3.4 Ancillary Treatment for Airway Management and Respiratory Support

Bulbar and respiratory paralysis often manifest in severe Cobra envenoming cases or patients undergoing delayed antivenom treatment. Therefore, antivenom therapy alone cannot be relied upon to manage these patients efficiently, and ancillary treatment involving airway management and respiratory support must be undertaken. The WHO has emphasized that although mechanical ventilation has been suggested for severe neurotoxic envenoming, several Cobra envenomed victims continue to succumb to death by asphyxiation each year because the attending physician assumes that antivenom administration alone can lead to recovery of the patients. Therefore, respiratory support must be promptly provided to patients diagnosed with bulbar and respiratory paralysis (WHO, 2016). The WHO outlines specific indications for respiratory support in envenomed patients, which include imminent respiratory arrest, neck muscle weakness with shallow respiration or paradoxical breathing, upper airway obstruction with stridor, Oxygen saturation (PaO_2) < 60 mm Hg, and blood gas measurement showing respiratory acidosis (hypoxia PaO_2 < 60 mm Hg with $PaCO_2$ > 45 mm Hg).

Artificial respiratory support can be provided to patients by non-invasive methods (Bag-mask/Bag-mask-valve ventilation/Ambu bag) or invasive ventilation (endotracheal tube and i-gel supraglottic airway). While non-invasive Ambu bags can provide initial respiratory assistance even at the primary healthcare centers, qualified physicians usually perform invasive methods on Cobra-envenomed patients with severe respiratory paralysis at district-level and tertiary hospitals.

In the Asian continent, only a few Cobra-envenomed patients who required invasive artificial respiratory support have been reported since the antivenom administration could reverse neurological disorders, thereby preventing the manifestation of severe respiratory paralysis. However, patients arriving late to the hospital and with severe envenoming episodes require invasive respiratory support. In a retrospective epidemiological study from Bangladesh, 11 out of 70 *N. kaouthia* envenomed victims (~15%) required artificial ventilation: 4 patients for 2 hours, 5 for up to 24 hours, 1 for 36 hours, and one for 7 days. Out of the 11 patients, 10 recovered after the respiratory support. However, the patient, under ventilation for 7 days, eventually succumbed to death (Faiz et al., 2017).

Similarly, approximately 26% of *N. kaouthia* envenomed patients (n = 35) reported in an epidemiological study from Thailand required artificial ventilation. Five of the nine patients recovered after the respiratory support, while four could not be saved (Buranasin, 1993). Interestingly, none of the *N. kaouthia* envenomed patients (n = 19) from Assam, India, required any ventilation support to recover (Kakati et al., 2023). However, ~50% of the *N. naja* envenomed victims (n = 41) from Maharashtra, India, and ~16% of *the N. naja* envenomed victims (n = 25) from Sri Lanka underwent early intubations and artificial respiration that helped to manage the neurological disorders in the patients successfully (Bawaskar and Bawaskar, 2004; Bawaskar et al., 2008; Kularatne et al., 2009). On the contrary, none of the victims (n = 15) of *N. mandalayensis* envenomation in Myanmar required assisted ventilation (Sai Sein Lin et al., 2020). Therefore, the prerequisite of respiratory support for Cobra envenomed victims in Asia varies from species to species and across geographic locations due to intra-specific venom variations. The attending physician must be aware of such discrepancies and adhere to the WHO indications for providing respiratory support to Cobra-envenomed victims when and wherever necessary.

7.3.5 Surgical Intervention (Fasciotomy and Debridement)

Besides neurological manifestations, envenomation in almost all Cobra species (except *N. philippinensis*) induces local tissue damage at the envenomed site. Although the patients may survive the envenomation effects, they must lead the rest of their lives with permanent disabilities resulting from local tissue damage. Therefore, treatment of local impacts of Cobra envenomation becomes very important to reduce the morbidities in envenomed victims. Management of tissue necrosis is usually done by surgical debridement, fasciotomy in case of compartmental syndrome, and amputation in case of gangrene of the affected toes or limbs. Debridement is performed if Frank's skin necrosis (demarcated, hypo-, or hyper-pigmented areas with an odor of putrefaction) is detected near the bite site. It involves surgical excision of unhealthy tissues from the site of necrosis to prevent the risk of anaerobic sepsis. Fasciotomy is usually advised in advanced cases of necrotic tissue damage that has developed into compartment syndrome (characterized by a build-up of excessive pressure in osteofascial compartments (a group of muscles and nerves wrapped in a layer of fascia), causing ischemia, thereby leading to muscle and nerve necrosis). This surgical procedure involves removing or excising the fascia (a sheath of stringy connective tissue beneath the skin) to ease pressure in the muscle compartment.

However, there has been a lot of debate on the criteria for the requirement of fasciotomy in snakebite patients since animal model studies have revealed that severe envenoming causes irreparable damage to the muscles, leading to compartment syndrome (Garfin et al., 1984). Therefore, fasciotomy may provide minimal benefits for managing advanced local tissue necrosis (Cumpston, 2011; Darracq et al., 2015). Moreover, the diagnosis of classical signs of compartment syndrome may be challenging. Therefore, the WHO recommends directly measuring the intra-compartmental pressure through a cannula introduced into the compartment. In concordance with general orthopedic practice, a fasciotomy (a procedure used to decompress acute compartment syndrome) is advised if the pressure readings exceed 40 mm Hg (WHO, 2016).

It has been observed that most *N. atra* envenomed cases (~60%) have to undergo surgical interventions to manage the local wound. A study involving 183 *N. atra* envenomed patients from Taiwan reported that surgical debridement was performed in 42.8% of the patients, while fasciotomy was required in 26.6% of the patients. Moreover, amputation of the affected toe or finger was necessary in 7 patients to manage the spread of tissue necrosis in the vicinity of the envenomed site (Mao et al., 2018). Similarly, surgical debridement was needed in 12 out of 70 (~17%) *N. kaouthia* envenomed patients from Bangladesh (Faiz et al., 2017). However, surgical intervention was not required to manage local tissue damage in *N. kaouthia* envenomed patients (n=19) from Assam, India (Kakati et al., 2023).

On the contrary, physicians had to perform debridement in ~20% of *N. naja* envenomed patients from Sri Lanka to manage local tissue damage (Kularatne et al., 2009). Therefore, the requirement for surgical intervention may also vary from species to species and across geographic locations. Nevertheless, the decision to perform surgical management of local tissue damage must be taken with utmost caution to avoid unnecessary and dangerous debilitation of the affected wound that may attract other secondary complications.

7.3.6 Clinical Management of Side-Effects of Antivenom Therapy

Although the administration of antivenom remains the only scientifically validated treatment for snake envenomation, these equine-derived preparations are associated with significant adverse side effects that may sometimes be life-threatening (de Silva et al., 2016; Malasit et al., 1986). Reactions to antivenoms can be classified as early anaphylactic reaction, late reaction (serum sickness), and pyrogenic reaction. Early reactions usually manifest within a few minutes and up to 180 minutes post antivenom administration and are characterized by itching, urticaria, intestinal colic, fever, vomiting, diarrhea, dry cough, and tachycardia. In addition, some patients may also suffer life-threatening anaphylactic reactions, including bronchospasm, hypotension, and angioedema. These early anaphylactic reactions are induced primarily due to the activation of the complement pathways by IgG aggregates or residual Fc regions in the antivenom (Garcia et al., 2002; Morais et al., 1994; Patra et al., 2021; WHO, 2016). Late reactions (serum sickness) in antivenom-treated patients can be observed on 1st to 20th day post-administration. Clinically, the patients manifest itchy urticarial, fever, vomiting, diarrhea, immune complex nephritis, mononeuritis multiplex, lymphadenopathy (swelling of lymph nodes), joint and muscle pain, and, in rare cases, encephalopathy (affecting the brain) (LoVecchio and McBride, 2003; Morais et al., 2012; Patra et al., 2021; WHO, 2010).

However, if the early reactions in patients are appropriately managed, they are less likely to develop delayed antivenom reactions. Antivenom-treated patients develop late reactions due to the generation of antibodies by the human immune system against the heterologous equine antivenom (foreign protein). These human antibodies against the horse proteins form soluble antigen–antibody complexes that induce numerous delayed antivenom reactions. Fever and rigors characterize pyrogenic responses, drop in blood pressure, and vasodilation, manifest within 1 to 2 hours post antivenom administration. These reactions are observed in patients if the antivenom is contaminated with pyrogens, most commonly bacterial lipopolysaccharides (Patra et al., 2021; WHO, 2016).

Management of antivenom reactions is utmost necessary for recovery of the envenomed victims, and if they are neglected, it may lead to life-threatening scenarios. According to the WHO, early anaphylactic reactions should be managed with an intramuscular administration of epinephrine (adrenaline) at an initial dose of 0.5 mg for adults and 0.01 mg/kg for children. If the adverse reactions persist or worsen, the initial dose can be repeated at 5–10 minutes. In addition, patients with bronchospasm should be treated with short-acting β2 agonist bronchodilators such as terbutaline or salbutamol via an oxygen-driven nebulizer and antihistamines such as chlorphenamine at doses of 10 mg for adults or 0.2 mg/kg for children via intravenous injection (WHO, 2016).

Furthermore, intravenous hydrocortisone administration (100 mg for adults or 2 mg/kg for children) can also help mitigate early antivenom reactions. However, it has a shorter half-life (1–2 hours) compared to chlorphenamine (12–40 hours) and is therefore unlikely to act for a longer duration. Late antivenom reactions are usually treated with oral administration of antihistamines (chlorphenamine) at 8 mg/day for adults and 0.25 mg/kg/day for children. The doses should be given at intervals of 6 hours and continued for five days. However, patients unresponsive to antihistamine therapy for 24–48 hours should be treated with prednisolone (corticosteroid) at doses of 20 mg/day for adults and 0.7 mg/kg/day for children given at intervals of 6 hours for 5–7 days. Pyrogenic reactions are managed conservatively by physically cooling the patient's body and oral administration of antipyretics (paracetamol) to alleviate the fever. However, intravenous fluids and adrenaline should be administered in patients suffering from severe hypotension and anaphylaxis (WHO, 2016).

The frequency of adverse effects of antivenom is highly variable. It is dependent on several factors, including antivenom format [IgG or $F(ab')_2$], specificity (MAV or PAV), and the amount of antivenom administered to the patients. For example, 81% of snakebite patients in Sri Lanka developed reactions against Indian polyvalent antivenom, of which 43% had severe reactions (Ariaratnam et al., 2001; de Silva et al., 2016).

Nevertheless, a more recent retrospective study from Sri Lanka on 398 antivenom-treated snakebite patients suggested that adverse reactions during antivenom treatment have significantly reduced to ~30–50% (Waiddyanatha et al., 2022). On the contrary, the same antivenom induced adverse reactions in ~5–55% of the patients in India, among which 10–15% suffered severe antivenom reactions (Cherian et al., 2013; Deshpande et al., 2013; Kakati et al., 2023; Menon et al., 2016; Srimannarayana et al., 2004). Notably, these figures represent cumulative data across several epidemiological studies that reported treatment of envenoming cases by all snake species prevalent in a particular geographical location. However, the exact frequency and severity of antivenom reactions exclusively in Cobra-envenomed patients treated with specific or paraspecific antivenom have not been reported discretely. Nevertheless, it is reasonable to believe that the frequency and severity of antivenom reactions in Cobra-envenomed patients will follow a similar pattern. A few studies briefly discussing these side effects in Cobra-envenomed victims have relied on the WHO guidelines to manage antivenom reactions successfully.

7.4 EFFICACY OF COMMERCIAL ANTIVENOMS IN REVERSAL OF NEUROTOXICITY AND LOCAL EFFECTS IN COBRA ENVENOMED VICTIMS

Cobra envenomation manifests two distinct pathophysiologies – acute neuroparalysis and local tissue damage, the former being life-threatening and the latter resulting in lifelong morbidities. Therefore, efficient hospital management of Cobra envenomed cases demands mitigating both these venom effects by antivenom therapy. While the neuroparalysis in envenomed victims is primarily caused by alpha neurotoxins (αNTX), the local effects are due to the presence of variable amounts of cytotoxins (CTXs) and phospholipase A_2 (PLA_2) in Cobra venoms. It has been observed in some epidemiological studies that administration of specific/paraspecific antivenoms can reverse the neurotoxic symptoms in Cobra-envenomed victims as early as within 30 minutes. However, the complete reversal of neuroparalysis might take longer, typically up to 24–48 hours, and may require artificial respiratory support (Bawaskar and Bawaskar, 2004; Bawaskar et al., 2008; Faiz et al., 2017; Kakati et al., 2023; Kularatne et al., 2009).

On the contrary, managing local effects is challenging and requires adequate ancillary treatment modalities. Along these lines, the inefficacy of even specific antivenoms in mitigating the cytotoxic effects of Cobra venoms in the Asian continent has been repeatedly highlighted (Faiz et al., 2017; Kularatne et al., 2009; Mao et al., 2018).

Moreover, it has also been emphasized that delayed antivenom treatment is one of the key factors contributing to antivenom ineffectiveness against local tissue damage (Madhushani et al., 2021). Once necrosis spreads rapidly, managing the damaged tissue becomes challenging, prompting debridement, fasciotomy, or even amputation of the affected extremities in severe circumstances. As discussed earlier, these surgical interventions can result in lifelong morbidities in the victims.

Two primary factors contribute to the challenge of treating tissue necrosis in Cobra-envenomed patients. First, the CTXs present in Cobra venoms are low molecular mass (~6 kDa) proteins. As a result, they do not elicit a robust immune response during the horse immunization for antivenom production. Therefore, the antivenoms contain relatively lower amounts of antibodies that target these CTXs when administered to the victims (Chanda et al., 2019; Chanda et al., 2018). Secondly, these toxins are firmly affixed to specific cell membrane components; most are retained at the envenomed site. As a result, unlike other Cobra venom toxins that enter the blood, vascular, and lymphatic systems for dissemination, CTXs are localized at the bite site (Lin et al., 2022). Thus, they are not available in circulation for neutralization by intravenously administered antivenom. This has been experimentally proved in a recent pilot study on four *N. atra* envenomed patients. The authors observed that considerable amounts of CTXs could be detected in the necrotic tissues and wound discharge fluids at the bite sites even after ten days of treatment with high doses of specific antivenom (Lin et al., 2022). Therefore, managing local tissue damage is challenging in Cobra-envenomed patients with conventional antivenom therapy. It necessitates the calls for alternative strategies for delivering antivenom locally to the envenomed sites for specific neutralization of CTXs. In this regard, topical antivenom or small molecule inhibitors (designed against Cobra venom CTXs) can be experimented with to combat local tissue damage in Cobra-envenomed victims.

7.5 CHALLENGES IN ANTIVENOM THERAPY AND SCOPE OF IMPROVEMENT FOR EFFICIENT MANAGEMENT OF COBRA ENVENOMATION

Safety and efficacy of antivenoms are the two most important factors determining the outcome of Cobra envenomation treatment. Although the exact frequency and severity of antivenom reactions in Cobra-envenomed patients are undetermined, the cumulative data across antivenom-administered cases suggest that many patients develop adverse side effects due to antivenom therapy (Ariaratnam et al., 2001; Cherian et al., 2013; de Silva et al., 2016; Deshpande et al., 2013; Kakati et al., 2023; Menon et al., 2016; Srimannarayana et al., 2004; Waiddyanatha et al., 2022). This grave concern is a hurdle for effective hospital management of snakebite victims and thus requires immediate attention. In the case of most antivenoms, the final product undergoes preliminary in-house quality assessment; however, a thorough quality control check to determine the batch-to-batch variation in safety and efficacy is neither performed by the manufacturing company nor other competent laboratories or regulatory authorities. Moreover, although the WHO has recommended several assays for the routine assessment of antivenom quality, safety, and efficacy (summarized in Table 7.2), they are usually not conducted comprehensively by all

TABLE 7.2
Parameters That Should Be Assessed in the Routine Quality Control of Antivenoms, According to the WHO Guidelines for the Production, Control, and Regulation of Snake Antivenom Immunoglobulins

Parameters	Testing Methods	Requirements
Appearance	Observation of color, cloudiness, and appearance of the powder (in case of freeze-dried preparations)	Compliance with the description of the marketing dossier
Solubility (freeze-dried preparations)	Addition of solvent and observation of time to reach solubility and appearance	The product should be completely dissolved within 10 min. The solution should not be cloudy.
Extractable volume	Extraction of the total volume from the vial or ampoule	Compliance with the indications of the label
Venom neutralizing potency	Determination of the Median Effective Dose (ED50), that is, the ability to neutralize the lethal effect of venom. Expressed as the mg venom neutralized per ml antivenom	Compliance with the specifications of ED50 for the product or with the requirements of national regulatory agencies
Osmolality	Determination of osmolality by accepted methods	At least 240 mOsm/kg
Identity test	Biological, physicochemical, or immunological tests	The identity of antivenom confirmed
Protein concentration	Kjeldahl, colorimetric, or spectrophotometric procedures	Not more than 10 g/dl
Purity and integrity of immunoglobulin	SDS-PAGE under reducing and non-reducing conditions	IgG or IgG fragments should comprise more than 90% of the proteins
Molecular size distribution	Size exclusion chromatography (FPLC or HPLC)	The product should consist mainly of monomeric IgG or IgG fragments
Test for pyrogen substances	Rabbit pyrogen testLimulus Amebocyte Lysate (LAL) test when validated and approved by the national regulatory agency	Accepted limits by the Pharmacopeia in use
Abnormal toxicity test	Intraperitoneal injection and observation of signs of toxicity (not required by some regulatory agencies)	No evidence of toxicity
Sterility test	Filtration through membranes, neutralization (when preservatives are used), and addition to culture media (trypticase soy broth and thioglycolate)	Absence of microbial growth
Concentration of sodium chloride and other excipients	Standard chemical methods	Specifications of Pharmacopeia and Regulatory Agencies

(*Continued*)

TABLE 7.2 (CONTINUED)
Parameters That Should Be Assessed in the Routine Quality Control of Antivenoms, According to the WHO Guidelines for the Production, Control, and Regulation of Snake Antivenom Immunoglobulins

Parameters	Testing Methods	Requirements
Determination of pH	By using a potentiometer (pH meter)	Specifications of Pharmacopeia and regulatory agencies (generally neutral pH)
Concentration of preservatives	Spectrophotometric or HPLC-based methods	Phenol: maximum 2.5 g/LCresol: maximum 3.5 g/L
Chemical agents used in plasma fractionation	Several methods, depending on the chemical agent to be quantified	Specifications of Pharmacopeia, national regulatory agencies, or manufacturers
Residual moisture (freeze-dried preparations)	Karl-Fischer titration gravimetric or thermo-gravimetric methods	Less than 3%

Source: Reproduced from Patra et al., 2021, Drug Testing and Analysis, 1–19.

antivenom manufacturers. Therefore, these factors increase the likelihood of the distribution of antivenoms that are unsafe and/or have sub-optimal efficacy (Patra et al., 2021). Thus, antivenom manufacturing companies must take utmost care to routinely determine the safety of the final products as per the recommendations of the WHO. Furthermore, the national regulatory authorities of the Asian countries must include the WHO-recommended guidelines for the production, control, and regulation of snake antivenoms in their framework and assess the set parameters in entirety before approving the antivenom for clinical use. This effort will significantly contribute to minimizing the adverse effects of antivenoms in the treated patients.

For the successful and effective management of Cobra envenomation, the existing antivenoms must be potent enough to reverse the local and systemic pathophysiologies without degenerating the plight of the patients. It is noteworthy to mention that although certain degree of immunorecognition and subsequent neutralization of Cobra venom toxins by commercial antivenoms have been demonstrated across several in vitro and in vivo studies (Chong et al., 2022; Wong et al., 2016), sufficient evidence from clinical settings is scarce since well-designed randomized clinical trials on snake antivenom are usually not performed and are considered unethical. As a result, the preclinical assessment of antivenom in mouse models is still viewed as the 'gold standard' assay for determining antivenom efficacy and safety. However, a fundamental constraint in such in vivo studies becomes relevant for managing Cobra-envenomed patients. The preclinical effectiveness of antivenoms is conveniently measured in terms of neutralizing venom-induced lethality caused by neuromuscular paralysis in the case of Cobra venoms.

However, neutralizing local tissue damage, the other essential pathophysiology of Cobra envenomation is usually not assessed in the preclinical assays. Therefore, the exact potency of commercial antivenoms against local effects remains largely unknown.

With the growing consensus among clinicians and scientists on the poor efficacy of antivenoms against local tissue damage, it is high time to develop alternative assays to assess the potency of commercial antivenoms against local effects, specifically those caused by Cobra venoms.

The other major challenge of the existing antivenom therapy is neutralizing low molecular mass (<15 kDa) toxins, constituting nearly 90% of total Cobra venom. Due to their smaller size, the low molecular mass protein exhibits poor antigenicity and thus cannot elicit a robust immune response in horses during antivenom production. Subsequently, the antivenoms do not contain sufficient antibodies against these low molecular mass toxins for efficient neutralization (Chanda et al., 2019; Chanda and Mukherjee, 2020b; Chanda et al., 2018; Deka et al., 2023; Vanuopadath et al., 2022). The problem is further aggravated when venoms from a single geographic location are used during immunization. Due to tremendous intra-species geographic venom variability, the resultant antivenom fails to perform efficiently at sites distant from the venom source location (Senji Laxme et al., 2021; Sintiprungrat et al., 2016).

The failure of antivenom becomes particularly evident in countries with large geographic areas, such as India. Therefore, numerous toxinologists have recommended that venoms from different geographic locations be pooled and used during antivenom production to account for the possible geographic venom variability. Alternatively, the government should support the development of region/country-specific antivenoms for efficient hospital management of snakebite victims. Along these lines, the production and clinical application of Polyvalent Anti-snake Venom Serum against venoms of *N. naja*, *B. caeruleus*, *D. russelii*, and *E. carinatus* by the National Institute of Health, Pakistan, is praiseworthy (Qureshi et al., 2013). The distribution of snake species in India and Pakistan is broadly similar, with the 'Big Four' species being dominant in both countries. Therefore, earlier, Pakistan imported PAVs manufactured in India to manage snake-envenomed patients. However, due to the existence of geographic venom variability and the resulting sub-optimal efficacy of Indian PAVs in Pakistan, they have now started producing Pakistan-specific PAVs using endogenous venom samples in the immunization mixture. A similar effort is being tried and tested for Sri Lanka. Premium Serums and Vaccines Pvt. Ltd., India, has produced a Sri Lankan-specific PAV exclusively for the island nation (Patra et al., 2021). However, the product has yet to be used in clinics.

7.6 ALTERNATIVE STRATEGIES TO COMMERCIAL ANTIVENOMS FOR TREATING COBRA BITES

While the failure of antivenoms due to geographic venom variability can be addressed by altering the immunization mixture, the poor neutralization of abundant low molecular mass Cobra toxins still poses an imminent threat to Cobra-envenomed patients, which requires meticulous scientific intervention. A simple but effective way of counteracting this problem might be supplementing the existing antivenoms with neutralizing antibodies explicitly raised against the low molecular mass Cobra toxins. However, this might significantly increase the cost of the final product, which may not be feasible in developing countries (Mukherjee and Mackessy, 2021). Therefore, alternative innovative methods must be explored to produce potent and safe antivenom with lesser manufacturing costs. In this regard, the development of camelid- and ovine-based systems has been proposed as cost-effective alternatives for antivenom production (Cook et al., 2010).

7.6.1 Development of Engineered Antibodies to Tackle Cobra Envenomation

We are thriving in an era of biotechnology research where recombinant DNA technology has propelled several drug development and manufacturing advancements. Therefore, developing phage-display libraries to produce recombinant humanized antibodies against snake venom toxins can be an attractive strategy for meeting the growing need for antivenoms in the Asian continent. Notably, since the antibodies are humanized, they are believed to elicit minimum to negligible serum reactions in administered patients (Laustsen et al., 2021; Nazari et al., 2020).

Another strategy can be to develop single-chain fragment variable antibodies (scFvs) for neutralizing snake venom toxins (Kazemi-Lomedasht et al., 2019; Liu et al., 2022). The scFvs are low molecular mass (~25–30 kDa) fragment antibodies that consist of the variable heavy and variable light chains connected by a linker, and they structurally resemble the fragment antigen-binding region (Fab) of a whole antibody. Notable advantages of using this format of the truncated antibody include higher tissue penetrability due to smaller size and absence of interference with the host's immune system (Eskafi et al., 2021; Mukherjee and Mackessy, 2021).

Apart from phage-display and scFvs, a few preliminary studies have also shown that it is possible to generate antibodies against synthetic custom-made peptides instead of whole venoms that recognize and neutralize the toxic effects of certain snake venom toxins. The most important advantage of these recombinant antivenoms is that they can be produced without snake venoms, which are required in large amounts for conventional antivenom production schemes and are often unavailable to antivenom manufacturers due to increasing demands (Laustsen et al., 2021; Sobrinho et al., 2019). Nevertheless, all recombinants mentioned above/custom antivenoms must be investigated thoroughly under in vitro and in vivo conditions to ensure their efficacy and safety. Moreover, their commercial viability must also be assessed if they are to be proposed for clinical applications (Mukherjee and Mackessy, 2021).

7.6.2 Natural Products for Cobra Bites: Can Phytochemicals Be Used as Axillary Treatment Along With Antivenom Therapy?

The challenges in the current antivenom therapy have also attracted scientists to explore natural products from plants and other sources that can mitigate the toxic effects of certain snake venom toxins with fewer side effects. The Asian continent is gifted with unprecedented floral diversity, and there is a widespread of medicinal plants that have been traditionally used to treat snakebites. Although a few studies have demonstrated Cobra venom-neutralizing activity of selected medicinal plants under in vitro and in vivo conditions (Table 7.3), medicinal plants are yet not approved for the treatment of Cobra-envenomed patients (Mukherjee and Mackessy, 2021; Puzari et al., 2022).

TABLE 7.3
A List of Bioactive Molecules Isolated From Traditional Medicinal Plants Showing Cobra Venom Inhibitory Activity

					Evidence			
Bioactive Ingredient	**Source**	**Molecular Weight (Da)**	**Cobra Species Tested**	**Cobra Venom Properties Inhibited**	**In silico**	**In vitro**	**In vivo**	**Ref**
2-hydroxy-4-methoxy benzoic acid	*Hemidesmus indicus* R. Br.	168.0	*N. kaouthia*	Lethality	No	Yes	Yes	Alam and Gomes (1998)
Lupeol acetate	*Hemidesmus indicus* R. Br.	468.8	*N. kaouthia*	PLA_2 activity, respiratory paralysis, cardiotoxicity and lethality	No	Yes	Yes	Chatterjee et al. (2006)
1-hydroxytetratriacontane-4-one	*Leucas aspera*	508.0	*N. naja*	Superoxide dismutase (SOD), catalase, and glutathione (GSH) activities and lethality	No	Yes	Yes	Venkatesan et al. (2014)
Stigmasterol	*Pluchea indica*	412.0	*N. kaouthia*	PLA_2 activity, cardiotoxicity, neurotoxicity, respiratory paralysis and lethality	No	Yes	Yes	Gomes et al. (2007)
β-sitosterol	*Pluchea indica*	414.0	*N. kaouthia*	PLA_2 activity, cardiotoxicity, neurotoxicity, respiratory paralysis and lethality	No	Yes	Yes	Gomes et al. (2007)
AIPLAI (*Azadirachta indica* PLA_2 inhibitor)	*Azadirachta indica*	1052.52	*N. naja, N. kaouthia*	PLA_2 activity	No	Yes	No	Mukherjee (2008)

Quercetin-3-O-α-rhamnoside	*Euphorbia hirta*	448.38	*N. naja*	Proteolytic, PLA_2, hemolytic, edema-reducing, hyaluronidase, lethal activities	Yes	Yes	Yes	Gopi et al. (2016), Samy et al. (2008)
Ikshusterol3-O-glucoside	*Clematis gouriana*	592.8	*N. naja*	Hemolytic and PLA_2 activities	Yes	Yes	No	Muthusamy et al. (2018)
WSG	*Withenia somnifera*	27,000.0	*N. naja*	Cytotoxicity, edema-inducing activity, myotoxicity	No	Yes	Yes	Deepa and Gowda (2002)
Turmerin	*Curcuma longa*	14,000.0	*N. naja*	Lethality and PLA_2 activity	No	Yes	Yes	Chethankumar et al. (2010)

A phytochemical viz. 2-hydroxy-4-methoxy benzoic acid (168 Da) isolated from the root extracts of the Indian medicinal plant 'sarsaparilla' (*Hemidesmus indicus* R. Br.) was one of the first natural products to demonstrate neutralization of lethality induced by Indian monocled Cobra *N. kaouthia* venom (Alam and Gomes, 1998). Combining this small phytochemical with anti-*Daboia russelii* rabbit antiserum at a 1:1 ratio demonstrated the inhibition of the lethality of *N. kaouthia* venom tested in male albino mice. However, the mechanism of action of 2-hydroxy-4-methoxy benzoic acid was not studied in detail. Moreover, the study did not explain how anti-*Daboia russelii* (a viperid) antibodies could inhibit lethality induced by *N. kaouthia* (an elapid) venom. Although some extent of cross-reactivity might be expected due to the presence of certain conserved venom protein classes across snake families, that may not be enough to neutralize the αNTXs, the primary lethality factor in *N. kaouthia* venom, since these neurotoxic proteins are usually not found in *D. russelii* venom. In any case, the exact mechanism of action of 2-hydroxy-4-methoxy benzoic acid is warranted.

A few years later, in 2006, another small molecule named Lupeol acetate (468.8 Da) from the root extracts of the same medicinal plant was reported to inhibit the PLA_2 activity, respiratory paralysis, cardiotoxicity, and lethality induced by Indian *N. kaouthia* venom in male albino mice (Chatterjee et al., 2006). However, similar to 2-hydroxy-4-methoxy benzoic acid, the exact molecular mechanism of Lupeol acetate was not investigated. Similarly, 1-hydroxytetratriacontane-4-one (508.0 Da), a triterpenoid isolated from *Leucas aspera* leaves, was also shown to inhibit the *N. naja* venom-induced lethality in Swiss albino mice (Venkatesan et al., 2014).

Plant-derived sterols such as stigmasterol and β-sitosterol isolated from root extracts of Indian camphorweed (*Pluchea indica*) have also been demonstrated to inhibit in vitro PLA_2 activity, and cardiotoxicity, neurotoxicity, respiratory paralysis, and lethality induced by Indian *N. kaouthia* venom in experimental animals. Moreover, a combination of stigmasterol and β-sitosterol at a 1:3 ratio enhanced the neutralization potency of commercial Indian polyvalent antivenom against *N. kaouthia* venom-induced lethality in Swiss albino mice (Gomes et al., 2007). Although β-sitosterol has been associated with antithrombotic activity (via thrombin inhibition) (Gogoi et al., 2018), the precise mechanism of these sterols in inhibiting diverse *N. kaouthia* venom activities needs further investigation. Neem (*Azadirachta indica*) is one of the critical Indian medicinal plants that has been reported to have various bioactivities. The methanolic extracts of neem leaves were shown to cause potent inhibition of PLA_2 activity of Indian *N. naja* and *N. kaouthia* venoms (Mukherjee et al., 2008). In addition, the active ingredient in the preparation (AIPLAI - *A. indica* PLA_2 inhibitor) could also inhibit tissue damaging, indirect hemolytic, edema, and anticoagulant activities of the above Cobra venom PLA_2s. Exploration of the mechanism of action of AIPLAI unveiled that it non-competitively inhibited the *N. kaouthia* PLA_2 enzymes (Mukherjee et al., 2008).

Apart from AIPLAI, Quercetin-3-O-α-rhamnoside (from *Euphorbia hirta*), Ikshusterol3-O-glucoside (from *Clematis gouriana* roots), WSG (*Withania somnifera* glycoprotein from *Withania somnifera*), and Turmerin (from *Curcuma longa*) have also been demonstrated to inhibit Cobra venom PLA_2s thereby neutralizing several PLA_2-associated pharmacological activities (Table 7.3; Figure 7.2) (Chethankumar et al., 2010; Deepa and Gowda, 2002; Gopi et al., 2016; Muthusamy et al., 2018; Samy et al., 2008).

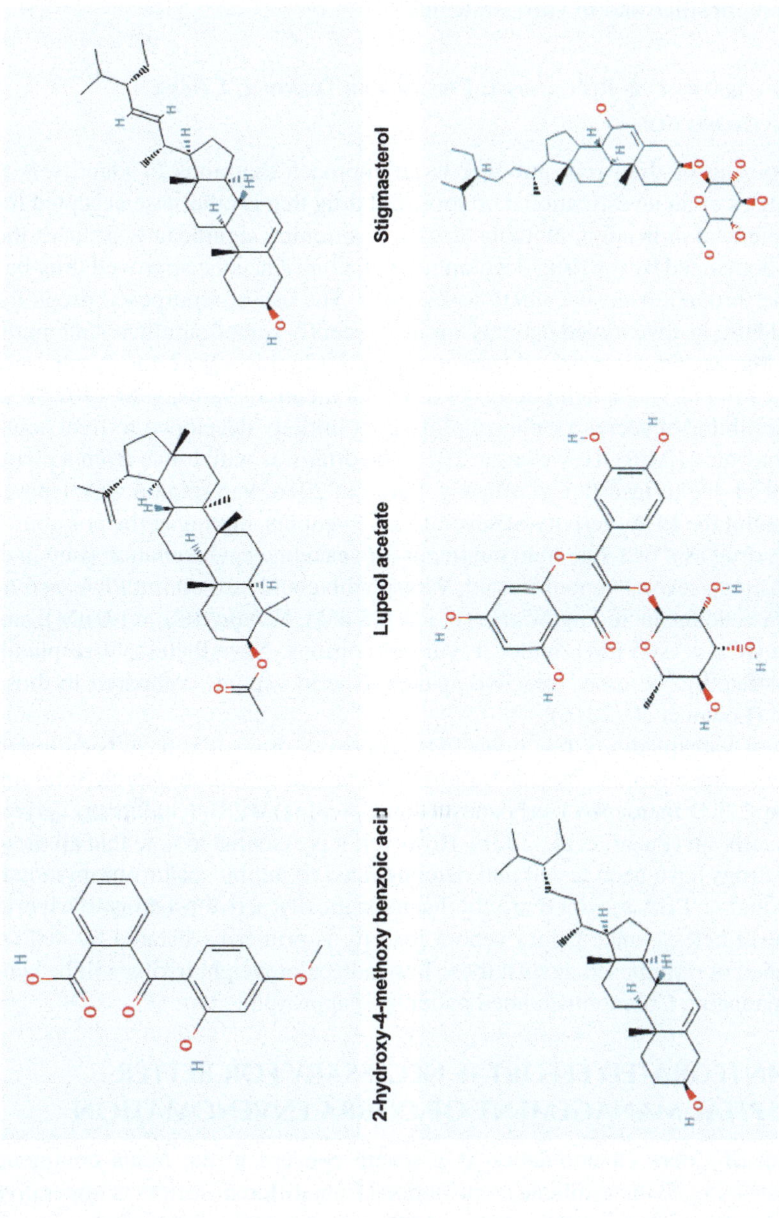

FIGURE 7.2 Chemical structures of the compounds that inhibit Cobra venom PLA_2 enzymes. The structures of the compounds have been retrieved from PubChem (https://pubchem.ncbi.nlm.nih.gov/).

Therefore, considering these promising findings, further research on systematic pharmacological screening of such medicinal plants for isolation and characterization of therapeutic molecules harboring anti-toxin activities should be encouraged and strengthened. Apart from natural compounds, small molecule inhibitors against snake venom toxins, which are either synthetic or occurring naturally, can also be screened using in silico and in vitro strategies.

7.6.3 Exploration of Repurposed Drugs for Treating Cobra Envenomation

Drug repurposing or drug repositioning is an approach that aims to identify new therapeutic uses of an investigational or approved drug that is otherwise accepted for a different medical indication. Notably, drug repositioning significantly reduces the time-frame demanded by the drug development pipeline since the approved drug has already gone through extensive safety assessment. The use of repurposed drugs for their possible use in envenomed patients has also recently gained significant attention from toxicologists and clinicians (Puzari et al., 2021). In this regard, the in vitro and preclinical testing of Varespladib deserves a special mention. Varespladib (402.38 g/mol) is an inhibitor of secretory phospholipase A_2 initially developed to treat acute coronary syndrome (ACS). However, in 2012, the drug was withdrawn from a phase III clinical trial due to insufficient efficacy. Later, in 2016, Varespladip was demonstrated to inhibit the PLA_2 activity of several snake venoms, hinting at the possibility of using this drug as a first-line field treatment for snakebite envenomation (Lewin et al., 2016). Among several venoms tested, Varespladib could also inhibit PLA_2 activities of Cobra venoms, including *N. atra* ($IC_{50} = 0.8$ nM), *N. naja* ($IC_{50} = 1.0$ nM), and *N. kaouthia* ($IC_{50} = 0.05$ nM), under in vitro conditions. Nevertheless, Varespladib was demonstrated to be more effective against Viperid venoms compared to those from Elapid (Lewin et al., 2016).

Apart from Varespladib, a few other FDA-approved drugs that have been tested against diverse snake venoms include Batimastat, Marimastat, Dimercaprol, Suramin, Quinolinone, 2,3-Dimercapto-1-propanesul-fonic Acid (DMPS), Citalopram derivative, and Disulfiram (Puzari et al., 2021). However, it is essential to note that all these repurposed drugs have been tested and demonstrated to inhibit snake venom metalloproteases and/or PLA_2s, which are the haemostatically active proteins of viperid venoms. Nevertheless, since Cobra venom toxicity is primarily dictated by 3FTxs, drugs designed or repurposed against these low molecular weight toxins will be beneficial for managing Cobra-envenomed patients, if approved.

7.7 AN INTEGRATED EFFORT IS NECESSARY FOR BETTER HOSPITAL MANAGEMENT OF COBRA ENVENOMATION

Management of Cobra envenomation is a severe concern in the Asian continent. Therefore, utilizing all the available contributions from different sectors is imperative to curb the burden of envenomation cases. In this regard, basic and clinical research on Cobra venoms and antivenoms is significant for understanding Cobra venom-induced pathophysiologies and the efficacy and safety of available antivenoms.

Notably, toxinologists, herpetologists, and clinicians are already conducting basic and clinical research on Cobra venoms and antivenoms in numerous laboratories of educational, research, and medical institutes in the Asian continent. Therefore, government funding agencies must support these efforts and try to improve the existing infrastructure of these laboratories so that they can pursue advanced venom research aimed at adequate clinical management of envenomed patients (Mukherjee and Mackessy, 2021).

Along these lines, basic research on the proteomic, biochemical, pharmacological, and immunological characterization of Cobra venoms, and genomic and venom gland transcriptomes of Cobra species are crucial for understanding Cobra venom action (Chanda and Mukherjee, 2020a; Kalita and Mukherjee, 2019; Mukherjee, 2020). Equally important are the epidemiological studies on Cobra envenomation, which can aid in developing strategies for distributing adequate antivenoms in high-risk geographic locations. In addition, the government funding agencies must also encourage research on quality control evaluation of commercial antivenoms and the development of improved and safer antivenoms, modified antivenoms, and identification of natural products as antidotes. Some active laboratories can be designated regional or national centers for research on snake venom and snakebite therapy, particularly on Cobra species, which may be worthy and fruitful for improving Cobra bite therapy.

Considering the multitude of priorities of a nation, it is plausible that government funds may not always be sufficient to propel the research on snakebites. Therefore, herpetologists, toxinologists, NGOs, and clinicians must also be aware of other international funding organizations and trusts such as the Bill & Melinda Gates Foundation (B&MGF), the Global Health Innovative Technologies Fund (GHIT), the Wellcome Trust, the World Health Organization, and the United Kingdom National Health Service (UK–NHS) that can be approached for funding opportunities on snake venom research (Mukherjee and Mackessy, 2021). Moreover, it is vital to establish national and international collaboration to share the available resources and experimental findings to strengthen Cobra envenomation research. The government should intervene and allocate dedicated funds for bilateral and multi-institutional projects to ensure successful cooperation. Furthermore, such international collaboration can also be encouraged by organizing international seminars, workshops, and CMEs (Continuing Medical Education) on snake venom research and antivenom therapy, where experts from different countries can share their experiences and deliberate on the existing bottlenecks of Cobra bite treatment.

The rising cost of antivenom is another issue requiring the government's attention. Snake envenomation primarily affects the lives of the poverty-stricken population, and most often, the victim is the family's sole income earner. Therefore, the government should try to include snake antivenoms in the National List of Essential Medicines, making them available either free of cost or at subsidized rates in rural healthcare centers and private hospitals. This effort will significantly reduce the financial burden on the victims and their families. Moreover, developing dedicated compensation schemes and insurance policies for the victims' families will support low-income families. Such strategies will encourage the snakebite victims and their families to rush to the nearest healthcare facility for appropriate treatment rather than deliberating on financial constraints and consulting traditional healers (Mukherjee and Mackessy, 2021).

Antivenom manufacturers' role is also crucial in effective hospital management of snake envenomation, including those by Cobras. An active collaboration between the antivenom manufacturers and the toxinologists must exist to identify the loopholes in the existing antivenom therapy that clinicians usually experience during the treatment. The toxinologists, in turn, must try to determine the exact problem in the prevalent antivenom therapy and produce possible solutions to overcome it. For example, there is a consensus among clinicians and toxinologists that local effects of Cobra envenoming caused by CTXs cannot be fully reversed by the existing antivenoms, which lead to lifelong morbidities in the envenomed patients. Notably, the diverse mechanisms of actions of CTXs are already established and linked to local tissue damage caused during a Cobra envenomation. Therefore, efforts must now be dedicated to finding solutions to improve the existing antivenom and introducing other treatment modalities to counter account for these toxic Cobra venom toxins. Such a concerted effort from toxinologists, antivenom manufacturers, and clinicians can help mitigate the harmful effects of Cobra venom-induced pathophysiologies and improve the victims' lives.

General public awareness and education on snakebites are also of paramount importance for prevention as well as treatment of Cobra envenomation (see Section 7.8). Active participation from government bodies and NGOs is necessary for promoting public awareness at the grassroots level. The government may provide the required financial assistance to the NGOs or educational institutes for conducting snakebite prevention programs at the snakebite hotspot localities. Moreover, awareness among school children can be created by introducing chapters on identifying non-venomous and venomous snakes and basic first-aid measures post-envenomation in their course curriculum. In addition, the government should also try to set up a 24 × 7 emergency snakebite helpline center to assist victims and their relatives with necessary action during an envenomation incident or encounter with a snake. Management of snakebites, including Cobra envenomation, has been a long-standing problem in the Asian continent (Mukherjee and Mackessy, 2021). However, we believe that such integrated efforts from all the stakeholders will significantly help reduce the incidence and severity of snakebites soon.

7.8 GENERAL PUBLIC AWARENESS AND EDUCATION: IMPORTANT ELEMENTS FOR CURBING SNAKE ENVENOMATION

Snake envenomation, including those by Cobras in the Asian continent, mainly occurs in rural areas with inadequate healthcare systems. Therefore, emphasis must be placed on preventing snakes by promoting public awareness and educating people about snake envenomation in high-risk communities. To begin with, organizing snakebite prevention programs by the government and non-government organizations (NGOs), or even educational institutes to educate the rural population on the adoption of logical measures for preventing an encounter with a venomous snake species can significantly help in bringing down the burden of snake envenomation.

Some standard preventive measures include the use of closed footwear and long trousers during walking or working in the agricultural field or forest areas, the use of a flashlight while walking outside during the night as well as sufficient illumination in and around the house, encouraging the rural public to use a mosquito net while sleeping and creating buffer zones between agricultural lands and inhabiting areas (Mukherjee and Mackessy, 2021).

While prevention is better than cure, it is practically impossible to eradicate snake envenoming. Therefore, if a snake envenoming has already occurred, the victim and the people near the incident must know how to manage the patient and move him appropriately to the nearest healthcare center for treatment. Therefore, they must be aware of the "Do's and Don'ts" during snake envenoming and identify snake species, if possible. The rural population must also be reassured that prompt antivenom administration is the only effective treatment against snake envenomation, and therefore, they should not waste time by consulting traditional healers. Conversely, instead of demonstrating magic tricks to treat the envenomed patients, conventional practitioners should be encouraged and trained to provide first-aid measures for envenomation and direct the patients to the nearest healthcare center for appropriate treatment. Public awareness and community education must be deployed using all the available means, including posters, newspapers, television, radio, social media, and smartphone apps. In addition, provisions should also be made to organize puppet shows and drama performances to gain a wider dispersal of the relevant education on the management of snake envenomation. It is noteworthy to mention that the success of this community education and awareness demands concerted efforts from individuals from the rural population and all the stakeholders, including government organizations, NGOs, international health organizations, toxinologists, herpetologists, clinicians, and antivenom manufacturers (Mukherjee and Mackessy, 2021).

7.9 CONCLUSION

The current primary treatment modality of Cobra venom in Asia includes intravenous administration of specific or paraspecific antivenoms. However, it is essential to note that no Cobra venom detection kits are currently available for precise and confirmed identification of Cobra envenomation, and treatment is primarily based on syndromic diagnosis. Along with antivenom treatment, severe neurotoxic envenoming cases require ancillary treatment for airway management and respiratory support. Moreover, patients with extensive local tissue damage may require surgical intervention by fasciotomy and debridement to remove the necrotic tissues and halt the spread of tissue necrosis. Although antivenom is the only WHO-approved treatment option for Cobra envenomation, there are concerns regarding the efficacy of these antivenoms and their adverse effects. Notably, poor immunorecognition of the major cobra venom toxins, including CTX, αNTX, and PLA_2, and geographical variation in Cobra venom composition can challenge the effective hospital management of Cobra-envenomed patients. These hurdles call for innovative antivenom design and production, exploring natural products that can inhibit Cobra venom toxin action, and repurposing possible FDA-approved drugs as complementary therapeutic approaches for treating Cobra envenomation. Cobra envenoming is a grave concern in the Asian continent. Still, it can be mitigated by a concerted effort from all the stakeholders, including government organizations, NGOs, international health organizations, toxinologists, herpetologists, clinicians, and antivenom manufacturers.

REFERENCES

Alam, M.I., Gomes, A., 1998. Adjuvant effects and antiserum action potentiation by a (herbal) compound 2-hydroxy-4-methoxy benzoic acid isolated from the root extract of the Indian medicinal plant 'sarsaparilla' (*Hemidesmus indicus* R. Br.). *Toxicon* 36, 1423–1431.

Ariaratnam, C.A., Sjostrom, L., Raziek, Z., Kularatne, S.A., Arachchi, R.W., Sheriff, M.H., Theakston, R.D., Warrell, D.A., 2001. An open, randomized comparative trial of two antivenoms for the treatment of envenoming by Sri Lankan Russell's viper (*Daboia russelii russelii*). *Transactions of the Royal Society of Tropical Medicine and Hygiene* 95, 74–80.

Attarde, S., Khochare, S., Iyer, A., Dam, P., Martin, G., Sunagar, K., 2021. Venomics of the enigmatic Andaman cobra (*Naja sagittifera*) and the preclinical failure of Indian antivenoms in Andaman and Nicobar Islands. *Frontiers in Pharmacology* 12, 768210.

Bawaskar, H.S., Bawaskar, P.H., 2004. Envenoming by the common krait (Bungarus caeruleus) and Asian cobra (*Naja naja*): clinical manifestations and their management in a rural setting. *Wilderness & Environmental Medicine* 15, 257–266.

Bawaskar, H.S., Bawaskar, P.H., Punde, D.P., Inamdar, M.K., Dongare, R.B., Bhoite, R.R., 2008. Profile of snakebite envenoming in rural Maharashtra, India. *The Journal of the Association of Physicians of India* 56, 88–95.

Buranasin, P., 1993. Snakebites at Maharat Nakhon Ratchasima regional hospital. *The Southeast Asian Journal of Tropical Medicine and Public Health* 24, 186–192.

Chanda, A., Kalita, B., Patra, A., Senevirathne, W., Mukherjee, A.K., 2019. Proteomic analysis and antivenomics study of Western India *Naja naja* venom: Correlation between venom composition and clinical manifestations of cobra bite in this region. *Expert Review of Proteomics* 16, 171–184.

Chanda, A., Mukherjee, A.K., 2020a. Mass spectrometric analysis to unravel the venom proteome composition of Indian snakes: opening new avenues in clinical research. *Expert Review of Proteomics* 17, 411–423.

Chanda, A., Mukherjee, A.K., 2020b. Quantitative proteomics to reveal the composition of Southern India spectacled cobra (*Naja naja*) venom and its immunological cross-reactivity towards commercial antivenom. *International Journal of Biological Macromolecules* 160, 224–232.

Chanda, A., Patra, A., Kalita, B., Mukherjee, A.K., 2018. Proteomics analysis to compare the venom composition between *Naja naja* and *Naja kaouthia* from the same geographical location of eastern India: Correlation with pathophysiology of envenomation and immunological cross-reactivity towards commercial polyantivenom. *Expert Review of Proteomics* 15, 949–961.

Chatterjee, I., Chakravarty, A.K., Gomes, A., 2006. *Daboia russellii* and *Naja kaouthia* venom neutralization by lupeol acetate isolated from the root extract of Indian sarsaparilla *Hemidesmus indicus* R.Br. *Journal of Ethnopharmacology* 106, 38–43.

Cherian, A.M., Girish, T.S., Jagannati, M., Lakshmi, M., 2013. High or low- a trial of low dose anti snake venom in the treatment of poisonous snakebites. *The Journal of the Association of Physicians of India* 61, 387–389, 396.

Chethankumar, M., Divakar, G., Jangid, M., 2010. Turmerin, a protein from *Curcuma longa* L. prevent oxidative organ damage against *Naja naja* venom phospholipase A2 in experimental animal. *Journal of Current Pharma Research* 3, 29–34.

Menon, J.C., Joseph, J.K., Jose, M.P., Dhananjaya, B.L., Oommen, O.V., 2016. Clinical profile and laboratory parameters in 1051 victims of snakebite from a single centre in Kerala, South India. *The Journal of the Association of Physicians of India* 64, 22–29.

Morais, J.F., de Freitas, M.C., Yamaguchi, I.K., dos Santos, M.C., da Silva, W.D., 1994. Snake antivenoms from hyperimmunized horses: comparison of the antivenom activity and biological properties of their whole IgG and F(ab')2 fragments. *Toxicon* 32, 725–734.

Morais, V., Berasain, P., Ifran, S., Carreira, S., Tortorella, M.N., Negrin, A., Massaldi, H., 2012. Humoral immune responses to venom and antivenom of patients bitten by *Bothrops* snakes. *Toxicon* 59, 315–319.

Mukherjee, A.K., 2020. Species-specific and geographical variation in venom composition of two major cobras in Indian subcontinent: Impact on polyvalent antivenom therapy. *Toxicon* 188, 150–158.

Mukherjee, A.K. (Ed.), 2021. Prevention and treatment of the "Big Four" snakebite in India. In *The 'Big Four' snakes of India: Venom composition, pharmacological properties and treatment of envenomation*. Springer, pp. 145–161.

Mukherjee, A.K., Doley, R., Saikia, D., 2008. Isolation of a snake venom phospholipase A2 (PLA2) inhibitor (AIPLAI) from leaves of *Azadirachta indica* (Neem): mechanism of PLA2 inhibition by AIPLAI in vitro condition. *Toxicon* 51, 1548–1553.

Mukherjee, A.K., Kalita, B., Dutta, S., Patra, A., Maiti, C.R., Punde, D., 2021. Snake envenomation: Therapy and challenges in India. In *Handbook of venoms and toxins of reptiles*. Editor, S.P. Mackessy. CRC Press, pp. 581–592.

Mukherjee, A.K., Mackessy, S.P., 2021. Prevention and improvement of clinical management of snakebite in Southern Asian countries: A proposed road map. *Toxicon* 200, 140–152.

Muthusamy, K., Chinnasamy, S., Nagarajan, S., Sivaraman, T., 2018. Computational and in vitro insights on snake venom phospholipase A2 inhibitor of phytocompound ikshusterol3-O-glucoside of *Clematis gouriana* Roxb. ex DC. *Journal of Biomolecular Structure and Dynamics* 36, 4197–4208.

Nazari, A., Samianifard, M., Rabie, H., Mirakabadi, A.Z., 2020. Recombinant antibodies against Iranian cobra venom as a new emerging therapy by phage display technology. *The Journal of Venomous Animals and Toxins Including Tropical Diseases* 26, e20190099.

Palasuberniam, P., Chan, Y.W., Tan, K.Y., Tan, C.H., 2021. Snake venom proteomics of samar cobra (*Naja samarensis*) from the Southern Philippines: Short alpha-neurotoxins as the dominant lethal component weakly cross-neutralized by the Philippine Cobra Antivenom. *Frontiers in Pharmacology* 12, 727756.

Patra, A., Herrera, M., Gutierrez, J.M., Mukherjee, A.K., 2021. The application of laboratory-based analytical tools and techniques for the quality assessment and improvement of commercial antivenoms used in the treatment of snakebite envenomation. *Drug Testing and Analysis* 13, 1471–1489.

Patra, A., Mukherjee, A.K., 2021. Assessment of snakebite burdens, clinical features of envenomation, and strategies to improve snakebite management in Vietnam. *Acta Tropica* 216, 105833.

Puzari, U., Fernandes, P.A., Mukherjee, A.K., 2021. Advances in the therapeutic application of small-molecule inhibitors and repurposed drugs against snakebite. *Journal of Medicinal Chemistry* 64, 13938–13979.

Puzari, U., Fernandes, P.A., Mukherjee, A.K., 2022. Pharmacological re-assessment of traditional medicinal plants-derived inhibitors as antidotes against snakebite envenoming: A critical review. *Journal of Ethnopharmacology* 292, 115208.

Puzari, U., Mukherjee, A.K., 2020. Recent developments in diagnostic tools and bioanalytical methods for analysis of snake venom: A critical review. *Analytica Chimica Acta* 1137, 208–224.

Qureshi, H., Alam, S.E., Mustufa, M.A., Nomani, N.K., Asnani, J.L., Sharif, M., 2013. Comparative cost and efficacy trial of Pakistani versus Indian anti snake venom. *The Journal of the Pakistan Medical Association* 63, 1129–1132.

Ranawaka, U.K., Lalloo, D.G., de Silva, H.J., 2013. Neurotoxicity in snakebite—The limits of our knowledge. *PLoS Neglected Tropical Diseases* 7, e2302.

Ratanabanangkoon, K., 2023. Polyvalent snake antivenoms: Production strategy and their therapeutic benefits. *Toxins* 15, 517.

Sai Sein Lin, O., Myat Thet, N., Khin Maung, G., Than, A., Mi Mi, K., Myat Myat, T., Myo, T., Pyae Phyo, A., Oakkar Kyaw, K., Aye Zarchi, S., Du Wun, M., Htay, A., O'Shea, M., Mahmood, M.A., Peh, C.A., White, J., Warrell, D.A., 2020. Clinical importance of the Mandalay spitting cobra (*Naja mandalayensis*) in Upper Myanmar – Bites, envenoming and ophthalmia. *Toxicon* 184, 39–47.

Samy, R.P., Thwin, M.M., Gopalakrishnakone, P., Ignacimuthu, S., 2008. Ethnobotanical survey of folk plants for the treatment of snakebites in Southern part of Tamilnadu, India. *Journal of Ethnopharmacology* 115, 302–312.

Seneviratne, S.L., Opanayaka, C.J., Ratnayake, N.S., Kumara, K.E., Sugathadasa, A.M., Weerasuriya, N., Wickrama, W.A., Gunatilake, S.B., de Silva, H.J., 2000. Use of antivenom serum in snake bite: a prospective study of hospital practice in the Gampaha district. *The Ceylon Medical Journal* 45, 65–68.

Senji Laxme, R.R., Khochare, S., Attarde, S., Suranse, V., Iyer, A., Casewell, N.R., Whitaker, R., Martin, G., Sunagar, K., 2021. Biogeographic venom variation in Russell's viper (*Daboia russelii*) and the preclinical inefficacy of antivenom therapy in snakebite hotspots. *PLoS Neglected Tropical Diseases* 15, e0009247.

Silva, A., Maduwage, K., Sedgwick, M., Pilapitiya, S., Weerawansa, P., Dahanayaka, N.J., Buckley, N.A., Siribaddana, S., Isbister, G.K., 2016. Neurotoxicity in Russell's viper (*Daboia russelii*) envenoming in Sri Lanka: a clinical and neurophysiological study. *Clinical Toxicology* 54, 411–419.

Sintiprungrat, K., Watcharatanyatip, K., Senevirathne, W.D., Chaisuriya, P., Chokchaichamnankit, D., Srisomsap, C., Ratanabanangkoon, K., 2016. A comparative study of venomics of *Naja naja* from India and Sri Lanka, clinical manifestations and antivenomics of an Indian polyspecific antivenom. *Journal of Proteomics* 132, 131–143.

Sobrinho, J.C., Francisco, A.F., Simoes-Silva, R., Kayano, A.M., Ruiz Diaz, J.J.A., Garay, A.F.G., Arruda, A., Ferreira, A.S., Santos, A.P.A., Luiz, M.B., Teles, C.B.G., Pereira, S.S., Zanchi, F.B., Calderon, L.A., Zuliani, J.P., Soares, A.M., 2019. Antimyotoxic activity of synthetic peptides derived from *Bothrops atrox* snake gamma phospholipase A2 inhibitor selected by virtual screening. *Current Topics in Medicinal Chemistry* 19, 1952–1961.

Srimannarayana, J., Dutta, T.K., Sahai, A., Badrinath, S., 2004. Rational use of anti-snake venom (ASV): trial of various regimens in hemotoxic snake envenomation. *The Journal of the Association of Physicians of India* 52, 788–793.

Suchithra, N., Pappachan, J., Sujathan, P., 2008. Snakebite envenoming in Kerala, South India: clinical profile and factors involved in adverse outcomes. *Emergency Medicine Journal* 25, 200–204.

Suntrarachun, S., Pakmanee, N., Tirawatnapong, T., Chanhome, L., Sitprija, V., 2001. Development of a polymerase chain reaction to distinguish monocellate cobra (*Naja khouthia*) bites from other common Thai snake species, using both venom extracts and bite-site swabs. *Toxicon* 39, 1087–1090.

Tan, C.H., Tan, K.Y., Wong, K.Y., Tan, N.H., Chong, H.P., 2022. Equatorial spitting cobra (*Naja sumatrana*) from Malaysia (Negeri Sembilan and Penang), Southern Thailand, and Sumatra: Comparative venom proteomics, immunoreactivity and cross-neutralization by antivenom. *Toxins* 14, 522.

Van Dong, L., Quyen, L. K., Eng, K.H., Gopalakrishnakone, P., 2003. Immunogenicity of venoms from four common snakes in the South of Vietnam and development of ELISA kit for venom detection. *Journal of Immunological Methods* 282, 13–31.